Engines, Energy, and Entropy

A Thermodynamics Primer

John B. Fenn

Yale University

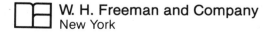

W. H. Freeman and Company
New York

Project Editor: *Patricia Brewer*
Copyeditor: *Stephen McElroy*
Designer: *Sharon Helen Smith*
Production Coordinator: *Bill Murdock*
Illustration Coordinator: *Richard Quiñones*
Artist: *John Johnson*
Compositor: *Bi-Comp, Inc.*
Printer and Binder: *The Maple-Vail Book Manufacturing Group*

Library of Congress Cataloging in Publication Data
Fenn, John B., 1917–
    Engines, energy, and entropy.
    Includes index.
    1. Thermodynamics. I. Title.
TJ265.F37   621.402′1   81-17305
ISBN 0-7167-1281-4   AACR2
ISBN 0-7167-1282-2 (pbk.)

Printed in the United States of America

5 6 7 8 9 10 VB 6 5 4 3 2 1 0 8 9 8

*To the memory
of the greatest thermodynamicist
of them all:
Josiah Willard Gibbs*

Contents

8 Two Laws From One Dilemma 129

9 All in a Nutshell 155

10 HER Has Much to Say 166

11 HER Under the Hood 191

Foreword

John Fenn has given us a remarkable book. His flair for apt whimsy can be seen immediately by glancing at the table of contents or the cartoons and verses. But prospective readers should not presume that they are being served a confection. Fenn's unorthodox exposition of thermodynamics pays tribute to both its majesty and its utility. Rather than yielding to sophistic pleas to avoid mathematics, he develops the requisite aspects of thermodynamic calculus in his own charming style.

The wondrous world of energy and entropy is not some academic fantasyland, yet it does resemble Marianne Moore's description: "Poetry is about imaginary gardens with real toads in them." Thermodynamics examines imaginary processes and thereby defines what is possible in our real physical world. These limitations are important, rigorous, and awesome in scope: They apply to galaxies, toads, and molecules. Like poetry, however, on first encounter thermodynamics often appears inscrutable, usually because the student gets lost in the gardens or is afraid of toads. The custom of blaming mathematics is silly. Hordes of students take courses in economics that employ the same level of calculus, although economics hardly ever gives reliable predictions and is never any fun.

John Fenn's didactic skill and wit make a profound science accessible to a wide audience. Key features of his approach are a host of canny examples and refreshing historical or cultural excursions. Beyond illuminating and enlivening the subject, the historical excursions reveal how science develops by fits and jerks, much as a toad travels. To many, this aspect may be just as instructive as the thermodynamic principles. Like art or music, science is a very human enterprise. Quite simplistic or downright wrong ideas often prove important by opening up new perspectives. Because the truth waits patiently to be discovered, primitive notions can be transmuted into elegant and durable concepts. By such an alchemical process thermodynamics slowly but surely emerged from puffs of steam to become one of the mightiest genies of science.

I thank John Fenn for creating a charming and insightful pathway into the gardens where this genie dwells.

December 1981

Dudley Herschbach
Harvard University

Preface

This book was born of two convictions: (1) To be considered liberally educated, one should have some understanding of the laws of thermodynamics; and (2) any reasonably intelligent person, even with little or no background in science and mathematics, can achieve such understanding with a modest expenditure of effort. The first of these propositions was asserted memorably and eloquently by C. P. Snow in his provocative *The Two Cultures*. I, too, believe that for liberally educated people, it is just as important to appreciate a Shakespeare play as it is to understand what turns the wheels of the vehicle that took them to the theater! The second of these convictions contradicts that bit of academic folklore which says that not even students of science and engineering can grasp the essence of thermodynamics until a second or third exposure. Certainly my own first encounter with the subject left me more confused than enlightened. But after 20 years of trying to teach and continuing to learn, I think I begin to understand why the folklore persists even as I become less persuaded of its accuracy.

The formalism and manipulations involved in thermodynamic analysis are simple—disarmingly and deceptively so. But the quantities to which the formalism relates are highly abstract. Some of them inhabit the observable and tangible world in more than one form and are often hard to identify. The difficulties of many students stem from the identification problem. They flounder in the process of translating a physical situation into the formalism, in abstracting and characterizing from what is observed the essential quantities to be manipulated. In the sense that the formalism can be regarded as a language for describing the real world, students falter in relating the objects and processes of that real world to the nouns and verbs of the language.

Of the conventional textbooks with which I am familiar, too many emphasize the formalism and its manipulations and neglect the translation problem. They grapple with the grammar and take the meaning for granted. This emphasis continues in the growing use of the axiomatic approach that reduces thermodynamics to an exercise in deducing a variety of formal statements from a few propositions that are simply asserted ab initio. In another popular trend one assumes the microscopic atomic–molecular model of matter and derives macroscopic thermodynamic statements from appropriate combinations of mechanics and statistics. These approaches have attractive features and should be part of a student's experience. Even so, to me they seem too contrived, too far removed from everyday experience with the visible, tangible, and audible world to form the basis of a working rapport between the student and the procedures of classical macroscopic thermodynamics that have been so effective in describing that world.

Historically, of course, what we call classical thermodynamics has its roots in the ability of perceptive minds to draw broad general conclusions from a relatively few specific observations. Its power and elegance attest to a triumph of induction. To be sure, deduction plays an important role in arriving at *specific* conclusions from general principles, but to understand and to use the general principles with confidence, to achieve facility in relating them to the physical world, I think one must retrace, at least vicariously, the inductive path that led the founders of the subject to their discoveries.

In putting this book together I have tried to emphasize and elucidate the translation process by appealing constantly to the reader's own experience and by relating it to the common experience of species *Homo sapiens*. I have long shared the view that the learning process of the individual closely parallels the learning process of the species. Thus, there is a fair amount of history stirred into this exposition. Most students appreciate and enjoy some historical perspective, but some have at first felt quite strongly that history and other humanistic touches have no place in a book on science and technology. Much to my surprise, most such negative reactions have come from students whose home base is in the arts, humanities, or social sciences! They cling to the image of completely impersonal and purely objective science and technology and seem distressed to discover that in means and ends scientific activities are very human endeavors. Those same students were uncomfortable with the informality and relaxed mood that I have deliberately tried to achieve. They were uneasy about the idea of Charlie. But eventually they relaxed and were able to enjoy learning with their shoes off, so to speak.

This book has been used (in manuscript form) in introductory courses in chemistry and physics. Because the standard textbooks in these courses must cover so many topics, they can spare space sufficient only for a barebones account of thermodynamic methods and their applications. For biology and geology students this treatment may well be the easiest introduction to a subject that is playing an increasingly important role in their disciplines. Students majoring in physical science or engineering, who take a course devoted entirely to thermodynamics, should find this book helpful, in part because its perspective and approach are quite different from those of their texts and in part because it answers some of the questions students hesitate to ask (and other books neglect to answer) because they seem too elementary. Despite its simplicity, or perhaps because of it, this book can be used as the primary text in a one-term course in thermodynamics for engineering students, especially freshmen or sophomores. If they understand clearly all the material covered, they will have as firm a grasp of the subject as most students who have used more conventional texts.

From these remarks, it has probably become quite clear that this book is not a conventional thermodynamics text. It does not pretend to be complete, rigorous, or elegant. Nor does it comprise a learning program—that is, an organized obstacle course, completion of which is supposed to ensure an ability to traverse similar but different terrains. Instead, its aim is to lead the reader to a comfortable understanding of the basic principles and methods of thermodynamics. Unabashedly elementary, it strives for simplicity with honesty.

October 1981   John B. Fenn

Acknowledgments

I am earnestly grateful to the many teachers and authors who have been the source for everything in these pages. Containing nothing new or original, this book is simply a repackaging of material that has been around for a long time, refined by many instructors and savored by many students—some fine old wine in a new bottle. Unfortunately, there are so many people from whom I have learned and relearned over the years that it is impossible for me to identify the origins of the ideas, perspectives, and flavors that may distinguish this book from the many others on the same subject. But I will record appreciation for a few with whom or with whose writings the memory of my encounters remains vivid.

Many years ago Benton Brooks Owen first exposed me to the subject, persuaded me of its power and importance, and introduced me to the book that still heads my list of favorites, the classic treatise by G. N. Lewis and Merle Randall (beautifully revised by K. S. Pitzer and Leo Brewer). Also on this book list are a text by J. H. Keenan, one by D. B. Spalding and E. H. Cole, Max Planck's famous monograph, the Columbia Lectures by Enrico Fermi, Joseph Kestin's "Course" and his translation of Ernst Schmidt's well-known exposition. Several essays by E. Mendoza and various articles in the *Encyclopedia Britannica* have seasoned my historical perspective. To all of these teachers and the host of others who must here be nameless—many, many thanks.

My appreciation goes to Karl Turekian, Juozas Vaisnys, Calvin Calmon, Walter Kauzmann, and Nicholas Delgass. At one time or another, each has provided encouragement and support when it was most needed. Then there are those comrades who finally launched this little vessel of words and phrases on its voyage to the printer. Dudley Herschbach, author of the delightful Foreword, engaged the interest and attention of Howard Boyer, who plotted the publication course at Freeman. Doug Vaughan made tight some leaky prose, and Patricia Brewer skillfully navigated the sometimes stormy passage to the printer. To all these kind friends and kindred spirits, my warmest thanks and appreciation. To my daughter Marianne goes the credit and my gratitude for bringing Charlie to life in her original sketches. Finally, to my endearing and enduring mate, my love.

J.B.F.

Engines,
Energy,
and Entropy

From BTU's and calories
Producing foot-pounds, ergs, and joules
Heat Engines must, to serve our ease
Obey inexorable rules.

1 In the Beginning

We start by introducing the very elementary concepts of *heat* and *work*. To emphasize how primitive they are, we will try to imagine how one of our forebears might have responded to his everyday experience. Necessarily, we will speculate a lot in this fanciful excursion to a time that provided no written records. We simply assume that a prototype early ancestor was blessed with the same set of senses and the same kind of brain that we enjoy, that his primary responses to various stimuli might have been as we will imagine them.

Then we hit a few high spots in the evolution that took twenty or so centuries to develop useful engines for obtaining work from heat. In retrospect, it is a wonder that so much time was required. In prospect, remarkable that it happened at all. In any event, the heat engine now plays a major role on civilization's stage. The rest of the book will identify and explain what the human race has learned about and from this leading performer and prime mover.

HOT AND COLD ARE DIFFERENT

Our prototype forebear never heard the term *thermodynamics* but Charlie the Caveman (as we'll call him) was an early student on the subject. He had a primitive but working knowledge of what our word *heat* means. Although his vocabulary was limited, he could probably have been about as effective in communicating the nature of heat as is today's product of ages of civilization and years of school. The word is hardly unambiguous. An ordinary dictionary gives ten meanings to the noun form and several more to the verb. Moreover, as we shall see, even in its original primitive sense, what we call ''heat'' puzzled some very bright minds for a very long time.

Let's get back to Charlie. He surely didn't understand all he knew about heat, but let's speculate about what he must have learned. Essentially, his

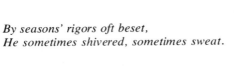

By seasons' rigors oft beset,
He sometimes shivered, sometimes sweat.

knowledge probably included what you and I have also been taught by firsthand experience. A rock in the shade felt different from a rock in the sunshine. The nearby shallow pond was more comfortable to wade in than the stream from the spring in the hillside. In the winter he shivered. In the summer he sweltered. The difference between hot and cold was meaningful to him because it had a direct impact on his senses. It was literally tangible.

Charlie also had some rudimentary conception of what a *thermal insulator* was, although he hardly called it that. Shade kept him cool on a hot summer day. Snow under leaves lasted well into spring. On the other hand, he could bury himself in dry leaves to keep warm. When he became more affluent, he wrapped up in furs. Sitting on a snow bank was more tolerable in a bearskin than in a bare skin.

After he tamed the fire that occasionally burned him out of his tree house, Charlie moved into his cave. Then his experience with heat was substantially broadened. Burning wood provided an intense heat source that was portable and controllable. It warmed things much faster than the sun. A rock close to a fire became too hot to touch. The first domestic hot water supply probably came about when a hot rock was dropped into the family water bucket. If anyone had asked him, Charlie would have said that yes, every time a hot object and a cold object were brought together the hot one cooled off and the cold one warmed up, but why such foolish questions about what everybody knows?

Charlie soon learned that objects in a fire or next to hot rocks underwent changes other than simply warming up. Wood would burn. Meat would cook. Ice and snow would melt. Water would boil or dry up. Clay would

Cooking food, beyond all question,
Spared our friend much indigestion.

bake hard. Stones might crack. The lessons on insulation also continued. If he wrapped his hands in green leaves, he could remove the well-done joint from the cooking fire without getting blisters. In his new bark-soled sandals he could walk on sun-baked rocks without getting a hotfoot.

GETTING HEAT FROM WORK

In due course Charlie learned something else that became very important to him. He found that he could generate heat by doing work. More precisely, he found that by working on an object he could get the same effect as when he brought it into contact with something hot. He could rub two sticks together and make them warm. In fact, if he rubbed them hard enough in the right way, he could start a fire. Of course, he couldn't have given a very good definition of work. Nor would he have had much to say about the difference between heating a stick by rubbing it and heating it by putting it into a fire. He was grateful for the rubbing process because it allowed him to start a new fire in case he got careless and let the old one go out.

Charlie was aware that the rubbing process took some effort on his part, that he had to exert himself. Indeed, human effort or exertion is the origin of the concept of work. In terms of firsthand experience, Charlie found, and everyone since has also learned, that the harder and farther one has to push or pull an object, the more tired one gets. Thus, the amount of work one does

To heat by work gave Charlie clout.
He could rekindle fires gone out.

is proportional both to distance and force. From such primitive experience comes the mechanical definition of amount of work as the product (by multiplication) of force and the distance through which it is exerted. In symbol shorthand:

$$W = F \times L \tag{1}$$

where W is the amount of work done on an object that is pushed (or pulled) by a force F through a distance L. Note that for work to be done, both F and L must be finite. You may exhaust yourself pushing on a brick wall, but unless you move it you do no work. Work may be what makes you tired but you can also become tired without doing any work of the kind we will be concerned with in this discourse.

There are a dozen meanings for the noun *work* and as many more for the verb that few of us ever stop to think about. We will adopt the simple definition of Equation (1). **Mechanical work is the product of force times distance.** Clearly, however, we need to flesh out this description. There must be some quantitative (numerical) characterization of both force and distance before we can meaningfully calculate how much work is done in any particular operation. Because the gravitational force on a unit mass (its weight) is pretty much the same everywhere on the earth's surface, and unchanging with time, one convenient and widely used measure of amount of work is determined by the distance through which a standard mass is raised against the force of gravity, for example, 1 foot-pound. Thus, 10 foot-pounds of work would be done if a 10-pound mass were raised 1 foot, a 5-pound mass were raised 2 feet, or a 1-pound mass were raised 10 feet. If you live in a country that has adopted the metric system, mass will be measured in grams or kilograms and distance in centimeters or meters. Whatever the units used for measurements, we will call any operation "work" if it results in, or *could result in*, the raising of a mass, that is, the exertion of a force (weight) through a distance (height). The meaning of *could* in this definition is a bit sticky sometimes. We will have more to say about it later on. Meanwhile, we are getting ahead of our story by a few centuries.

To Charlie, the main measure of how much work he did was how tired he got. For our present interest the importance of what he discovered and comprehended was that he could make something hot by working on it with his muscles. Unless we become Boy or Girl Scouts, most of us don't think much about starting fires by doing work. But we all soon learn that frictional work from sanding a piece of wood, stopping a car with brakes, or letting a bearing get dry can result in what we loosely call the generation of heat.

What we really mean is that the wood, the brake linings, or the bearings heat up just as though we had brought them into contact with a hot object.

HEAT AND WORK ARE DIFFERENT

We are going to have to be a bit fussy about distinguishing between what we call heat and what we call work. Here for clarification and future reference is our definition of **heat**:

> **Heat is the interaction (what happens) between a hot object and a cold object that are in contact with each other.**

Usually the hot object gets cooler, and the cool object gets hotter until they both have the same hotness. When no further heat interaction occurs, the two objects are in **thermal equilibrium.** (The word *equilibrium* comes from the Latin words *aequus* for "equal" and *libra* for "balance." It describes a condition or state of rest in a system, or between a system and its surroundings, when no changes are possible because all forces are in exact balance.) Hotness is the driving force that causes the interaction we call heat. When two objects are in thermal equilibrium, they have the same hotness.

Sometimes, instead of a change in "hotness," the result of a heat interaction can be a change in phase, like the melting of ice (solid phase to liquid phase) or the vaporization of water (liquid phase to vapor phase). Heat interactions between objects can be slowed down or inhibited by wood, cloth, fur, feathers, and the like. We call these materials *thermal insulators*. If we are ever in doubt as to whether an interaction between two objects is heat, we can make a test by repeating the operation with an insulator between them. If the observable change in either object is retarded or slowed down by the insulator, the interaction involves heat.

So much for heat, what about work? Here goes:

> **Work is any interaction that is not heat.**

Some people will not like that definition, but it's not a bad one. Thermodynamics classifies all interactions (other than a transfer of matter) as either *heat* or *work*. Consequently, an operational definition of either one defines the other by exclusion.

Of course, to understand either definition, we need a precise understanding of interaction. There is an **interaction** between a system (an object or collection of objects) and its surroundings (or another system) when there is a correlation, or correspondence, between an *observable change* in the system and one in its surroundings or in the other system. We may not understand the nature of the interaction or perceive the mechanism by which it occurs, but if there is a consistent correspondence between a change in a system and a change in its surroundings, we insist as an article of faith that there is an interaction between them.

We can't see, smell, hear, or feel the force field between a magnet and a compass needle, but we note that every time the magnet moves the needle moves. Therefore, there must be an interaction, and we invent the magnetic field to explain it. As a matter of fact, we don't directly observe what is going on when a hot and cold object are brought together. We observe the *consequence* of what is going on, that is, that the hot object cools off and the cold one warms up. We call what is going on, the interaction, heat. If we have doubts in a particular case, we reassure ourselves by the thermal insulation test. That is, we put a piece of thermal insulation between the objects and determine whether the rate of change in each one is affected. If it is slowed down, the interaction is heat. If there is no effect, the interaction is work.

More positive definitions of work are possible. Many teachers and authors prefer this one, set forth in 1873 by Josiah Willard Gibbs, a once obscure Yale professor now widely revered for his contributions to thermodynamics. Somewhat reworded, it says:

Work is any interaction between a system and its surroundings that has or *could* have as a sole effect in either the system or the surroundings the raising of a weight.

If the weight goes up in the surroundings (or could have gone up), the system is said to have performed positive work on the surroundings. If the weight has, or could have, gone up in the system, the system is said to have performed negative work on the surroundings, or the surroundings are said to have done positive work on the system. The problem with this definition is the phrase "could have as a sole effect." It's not always easy to decide what *could* have been. An observer would have to be omniscient to make that decision in every possible case. Let's consider an example that illustrates why it is sometimes easy to be confused about what *could* happen.

Suppose we hook up an electric battery to a switch and a hot plate in series. A container of water is on the hot plate (see Figure 1-1). The *system* (object with which we are concerned) is the battery. To identify its relation to the *surroundings* (rest of the world), a dotted line representing the *boundary* has been drawn around the battery. Now we throw the switch. The battery

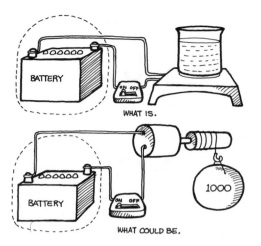

Figure 1-1 Things are not always what they seem to be. An interaction that *could* raise a weight (as a sole effect) is work, no matter what it actually *does*.

discharges through the hot plate and the water gets hot. Has the interaction between the battery and its surroundings been heat or work? The answer, which surprises some people, is work! Why? The battery could have been discharged through an electric motor that turned a windlass that raised a weight. With sufficient care we could have reduced the bearing friction in motor and windlass to a negligible level. The motor could have had super-conducting windings with no electrical resistance and no warming up would occur. If we exercised enough skill, we could have created a situation in which the *only* thing that happened in the surroundings (outside the boundary) was the raising of the weight. In that circumstance the interaction would qualify as work according to our positive definition.

But suppose we were able to take the battery, resistance heater, and bucket of water back a few millennia and let Charlie have a go at the question. If we had given him the weight-raising definition of work, he would have been up a tree because he'd never have heard of an electric motor. However, he could easily use our definition of heat to make a correct decision. He could wrap a bearskin around the battery and find empirically that it made absolutely no difference in the rate at which the water in the bucket got hot. Therefore, the interaction could not have been heat. Thus, it must have been work. Of course, since the wires have to pass through the bearskin, Charlie must ask whether there might have been enough heat interaction along the wires to warm the hot plate and the water. He could eliminate this possibility by passing the wires through a hot-water bath and then an ice-water bath on the way to the hot plate, or by varying the length of the wires. He would find no difference in heating rate. Because all such variations

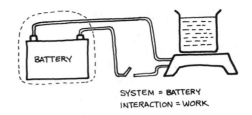

Figure 1-2 A lot depends on where a system's boundary is drawn. If the boundary included everything that changed, there would have been no interaction, neither heat nor work!

SYSTEM = BATTERY
INTERACTION = WORK

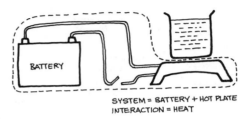

SYSTEM = BATTERY + HOT PLATE
INTERACTION = HEAT

would cause rate changes in cases where he could be sure an interaction is heat, he could reasonably conclude that this time it must not be heat.

Note very carefully that we are talking about what happened between the *battery* and the surroundings. If we had defined the system to include the hot plate and the battery, excluding the water, that is, if we had drawn the boundary so that it passed *between* the hot plate and the container of water, as in Figure 1-2, then the interaction across the boundary would have indeed been heat. Charlie's bearskin would have slowed it down. We must be very careful in making statements because what happens often depends upon how we define things, especially the system.

Even though this insulator test for identifying a heat interaction is simple and appealing, there are situations in which it can be misleading. Materials that inhibit a heat interaction might also interfere with some kinds of work interactions. Electromagnetic waves crossing the boundary of a system represent a case in point. In the familiar case of radiant heat exchange such waves result in what is clearly a heat interaction. On the other hand, the nonthermal coherent radiation of radio waves and laser beams does not ordinarily run motors that lift weights, but in principle it could. Such radiation is attenuated or reflected by the same kinds of mirror surfaces that provide effective insulation against purely thermal radiative heat transfer. In sum, like the Gibbs definition of work, the simple insulator test for heat may require more than a primitive understanding and knowledge to be useful in some situations.

ENTER A GREEK, HERO

In speculating about Charlie the Caveman's experience, we concluded that in the sense of perception and use he knew much of what you and I know about heat. The one great gap in his knowledge was the converse of the work-to-heat process. He knew that work could make a cold object hot. What he didn't know was that making a hot object cold could result in the performance of work. It can be argued that learning to kindle fire at will was probably the greatest single forward leap in the long journey from early animal existence to modern creature comfort. But the discovery that heat could be used to reduce human toil by performing work was the basic step in developing the technology that has been mankind's mixed blessing for the last few centuries. That discovery was a long time coming. Hero of Alexandria, the Greek scientist and mathematician, is credited with inventing a steam engine somewhere between 150 BC and AD 150.

Hero's aeolipile (Figure 1-3) is the grandfather of the turbine and the rocket. It was simply a hollow sphere free to rotate on the trunnions that supported it. Water was heated in a boiler, and the steam was introduced through a support member and exhausted through a pair of bent tubes attached to the sphere's periphery. The reactive force due to the ejected steam caused the sphere to rotate in much the same way as a fireworks pinwheel. Like the pinwheel, Hero's engine was more toy than labor-saving device. But it did demonstrate that a heat interaction between a fire and a boiler could result in the performance of mechanical work. Hero's description is the first clear record of a device that could bring about the heat-to-work process by design, although there is some evidence that priests used the motive power of steam under pressure to open doors and make idols move.

Of course, you should realize that for many centuries before Hero people had used the motive power of heat without really comprehending what they were doing. When someone first floated downstream on a log, he was taking advantage of the gravity-induced flow of water from the mountains to the sea. The water arrived at the upper mountain slopes having been evaporated

Figure 1-3   Heat to work. Hero's aeolipile (literally, wind ball), probably the world's first reaction turbine, may have never left the drawing board.

from the sea by the sun's heat and then condensed into rain or snow that was later melted, also by the sun. When the log evolved into a sailboat, the force exerted by the wind provided motion and thus did work. Wind is also a consequence of the sun's heat interaction with the atmosphere. Windmills as stationary sources of mechanical work were in wide use in the Low Countries in the twelfth century, long before serious development of the steam engine. But rafts and sailboats, windmills and water wheels, powered though they are by the sun's heat, did not result from controlled application of the ability of heat sources to provide work. We must credit Hero with the realization that work could be obtained at will from a heat interaction.

FIRST CAME GUN POWER

Though Hero's engine was an epochal invention, it hardly signaled the widespread replacement of muscle power by boiler–firebox combinations. Many years were to pass before really useful machines were developed, and they were hardly the direct descendants of Hero's aeolipile. The first applications of heat engines were as weapons of war—the rocket and the cannon. In the annals of Western civilization, recipes for making gunpowder from sulfur, charcoal, and saltpeter (potassium nitrate) were first set forth by the English friar Roger Bacon sometime before the middle of the thirteenth century. Bacon speculated about gunpowder's military applications but apparently did not contemplate its use as a propellant. Another ecclesiastic, the German monk Schwartz, apparently built the first cannon in AD 1313. By 1453 the art had so advanced that Mohammed II could order the construction of his famous Great Gun. Made of brass, it was 17 feet long and weighed almost 19 tons. It fired a 600-pound stone ball 25 inches in diameter for a distance of a mile!

It seems likely that Bacon learned about gunpowder from the Chinese, who had probably discovered saltpeter some centuries BC and used it in incendiary compositions for fire arrows. By AD 1225 they had developed

*'Tis sad that first into production
Oft are engines of destruction.*

the self-propelled rocket, which found its way as far east as Cologne by 1258. Superseded by guns, the rocket played only a minor military role until World War II. Now it is a major component of modern ordnance. More recently it has taken astronauts to the moon and remains our only practical means of locomotion in space.

REENTER STEAM

With heat engines, as with many other kinds of machines, military use preceded civil applications. The harnessing of heat as a motive power for constructive work did not really begin in earnest until the eighteenth century. Until then, all goods production in England and on the continent was literally by manufacture (from the Latin *manu* meaning "by hand" and *factus, "made"*). Motive power came from muscle. But the smelting of iron and other ores depended upon the heat (and the chemical reducing power) generated by burning charcoal, which was made from wood. The forests of England were so laid waste that the king finally prohibited further cutting of trees for the production of charcoal. This early act of conservation was not based on an esthetic or ecological concern for the environment—the navy and merchant marine of this sea-faring nation were becoming alarmed at the prospect of no wood to build ships!

When the charcoal supplies were cut off, it became necessary to go after coal. There were great reserves of this fuel, but they were under the water table. Horse-powered pumps could not cope with the flow of water into mines. So necessity resulted in the invention of a useful steam engine. During the centuries since Hero, there had been some desultory attempts to use steam to pump water. But it was not until 1698 that Thomas Savery, a mine owner from Cornwall, patented a steam-actuated water pump. Steam under pressure was introduced to a chamber containing water (a in Figure 1-4) until it forced the water out through valve b. With valve c closed, cold water was then sprayed on the outer surface of the chamber to condense the steam. The resulting vacuum sucked water into the chamber through valve d, and the cycle was ready to be repeated. This engine–pump worked, but it could raise water only a limited distance. And its consumption of fuel was prodigious. Today the principle survives in a coffeemaker reputed to provide a most savory brew.

The next important step in the development of steam engines happened around 1712 and is usually attributed to another British Thomas. Instead of letting steam push directly on the water surface, Thomas Newcomen, an iron monger who had worked with Savery, built an engine in which steam

Figure 1-4 Savery's steam pump. Patented in 1698, it never lifted much water.

pushed on a piston in a cylinder. This idea grew out of some correspondence between Newcomen and Robert Hooke, the English physicist. It had earlier been proposed and analyzed by Denis Papin, a French physicist who had been a student of Christian Huygens, the great Dutch physicist. Papin was somewhat distracted from his penchant for pure pistons after he was sent a description of Savery's pump by Wilhelm Leibniz, the German mathematician and philosopher who invented calculus. In a logical but fruitless undertaking, Papin designed an engine in which a piston-like structure separated the steam and water interfaces and inhibited the condensation of the steam on the water surface, a problem that plagued Savery's engine. This historical aside is simply a reminder that interest in steam engines was both international and interdisciplinary from the beginning.

In Newcomen's engine steam under pressure from a boiler pushed a piston to the top of a vertical cylinder (Figure 1-5). The steam was then condensed by injecting water into the cylinder. The resulting vacuum

Figure 1-5 The beginning of the piston age. In Newcomen's engine, steam pushed the piston up, but atmospheric pressure pushed it down as the steam condensed in the cylinder to form a vacuum.

Figure 1-6 Watt's incorporation of a separate steam condenser made possible the very rapid expansion of steam-engine use.

allowed atmospheric pressure to push the piston to the bottom of the cylinder. Newcomen engines were used almost exclusively for pumping water. When power was needed to turn other machinery, a Newcomen engine would pump water to a level from which it could fall through a water wheel! (This practice has had a reincarnation. Nowadays, some electric generating plants use off-peak power to pump water to elevated reservoirs. During peak demand, the water runs downhill through hydraulic turbines whose output supplements the capacity of the steam plant.) In spite of its wide use, the Newcomen engine left much to be desired. It was large, ungainly, and had an extravagant appetite for fuel.

In 1763 the Scottish instrument maker James Watt became disturbed by the Newcomen engine's great waste of heat through the alternate warming and cooling of the cylinder. He introduced the idea of a separate condenser that would remain cool yet communicate with the cylinder by a valve opened at a propitious point in the cycle (Figure 1-6). Thus, the cylinder could be kept warm all the time—in fact, could be insulated to keep from losing heat to the surroundings. Watt's condenser greatly cut fuel consumption and increased efficiency. It marked the real beginning of the steam engine age, which we are still very much in.

Over and above its stationary use, the reciprocating steam engine, based on a piston moving back and forth in a cylinder, provided the first self-contained locomotive power and led to the development of the steamship and the railroad. Its mechanical descendants in the form of the reciprocating internal combustion engine still run railroads, automobiles, some ships, and some airplanes as well as lawn mowers, chainsaws, and snow blowers. More recently the piston and cylinder have given way to a closer cousin of Hero's aeolipile, the turbine (Figure 1-7). Steam turbines ac-

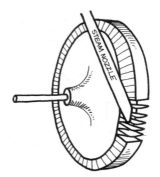

Figure 1-7 DeLaval's "steam windmill," one of the first impulse turbines.

count for most of the world's electric power. Gas turbines are widely used in aircraft propulsion. Some engineers believe they will ultimately displace the piston engine in automobiles as well. In a turbine the working fluid, steam or gas initially at high temperature and pressure, expands through a nozzle into a low-pressure region to form a high-velocity stream that impinges upon blades, or "buckets," mounted on a rotor. The action is similar to that of a windmill. Because it has fewer moving parts, no valves, and continuous unidirectional motion, the turbine requires less maintenance and can operate at higher speeds and with less heat loss than a piston engine. Consequently, its efficiency and power output per unit weight and size are much higher than for piston engines.

The development that began with Hero is of more than just historical interest. The heat engine is still very much with us. It is the source of most of the world's motive power. The only means of performing mechanical work that do not involve a heat-engine cycle are living muscle, the tides, and some esoteric devices that at very low efficiencies and in minuscule amounts can convert radioactivity directly into electric current. You might be interested in some figures. In 1979 about 2319 billion kilowatt-hours of electric power were generated in the United States alone. Of this total 12 percent came from falling water. The rest came from heat obtained by the combustion of fuel—48 percent from coal, 15 percent from gas, 13 percent from oil, and 11 percent from "burning" uranium. In terms of heat released in *all* applications, the combustion of coal, oil, and gas resulted in about 72×10^{15} BTU (British Thermal Unit, a unit of heat interaction; see Appendix I for a discussion of various units and measures). That much heat could make hot tea out of all the water that flowed over Niagara Falls that year.

Meanwhile, in 1979 automobiles and trucks consumed about 131 billion gallons of fuel equivalent to about 16×10^{15} BTU of combustion heat, somewhat more than half as much as was released in fireboxes under boilers. Central station generating plants run practically all the time. The average automobile is used only a few hours a day at most. All the automobile engines in the country running at the same time would generate 2.6×10^{10} horsepower, 33 times as much as all the generating stations. If all this power were put to work pumping sea water across the Isthmus of Panama, in one day the level of the Atlantic Ocean would drop over five feet while the level of the Pacific would rise about two and a half!

All of this activity by heat engines. No wonder that mechanical engineers call them *prime movers*. Surely we must try to understand these machines that play such a major role in our lives.

IN RETROSPECT

This brief account has collapsed a few million years of the history of heat into a few pages. These statements include what's worth remembering:

1. Some objects feel hot and some feel cold.

2. If a hot object is put in contact with a cold one, the hot object cools off and the cold one warms up. Finally, they reach a state of *thermal equilibrium* in which they both feel equally hot (or cold). We call what happens between the two objects *heat,* more properly, a *heat interaction.*

3. When there is a consistent correspondence between an observable change in a system and an observable change in its surroundings (or in another system), we say an *interaction* has occurred between a system and its surroundings (or the other system).

4. In addition to changing hotness, heat interactions can also melt ice, boil water, wilt leaves, cook meat, dry clothes, bake clay, and so on.

5. Many materials, for example, bark, fur, wood, and feathers slow down or inhibit heat interactions. We call these materials *thermal insulators.* If an insulator slows down the rate at which two objects observably interact, the interaction is heat. If the insulator has no effect, the interaction is work.

6. *Work* is any interaction that is not heat, in other words, any interaction not inhibited by a thermal insulator. Work can always be reduced to the raising of a weight as its *sole* consequence. Work interactions can have the same result as heat interactions, for example, making objects hotter.

7. Heat interactions can cause work interactions. Devices that convert heat into work are known as *heat engines*. They are the most important members of that family of machines known as *prime movers,* which are original sources of motive power.

WORKOUT ON WORK

1. Look up the words *heat* and *work* in an unabridged dictionary. How many distinguishably different meanings can you find for the verb and noun forms of each word?

2. List at least half a dozen phenomena, different in kind, that you have observed and for which you could empirically conclude that an interaction had occurred between two systems. For example, every time you open the refrigerator door, the inside light comes on. Therefore, there is an interaction between the door and the light. (Incidentally, how could you easily satisfy yourself that the light is off when the door is closed?)

3. Equation (1) says that $W = F \times L$ where W is the amount of work performed when force F acts over a displacement L. A barrel of oil comprises 42 gallons. Assume that oil weighs 8 lb/gal, and compute the amount of work in foot-pounds required to raise 1 barrel of oil to the surface of the earth from the bottom of a well 10,000 feet deep.

4. A typical American automobile engine can generate 200 horsepower. One horsepower is 33,000 ft lb/min. On a typical day the rate of flow of water over Niagara Falls is about 200,000 ft³/sec. The overall drop through the rapids and over the falls is 315 feet. Assume that water weighs 62.4 lb/ft³. How many automobile engines would it take to pump water to the top of the upper rapids as fast as it flows down?

5. Give as many examples as you can think of that represent processes in which work interactions have the same or nearly the same result as a heat interaction would have.

6. List all of the different kinds of prime movers that you have seen, read about, or can imagine.

7. Identify the different kinds of heat engines you have encountered, directly or indirectly.

2 · How Hot Is Hot?

One essential feature of both science and technology is quantitative description. To become appropriate subjects of scientific discourse, heat phenomena need to be cast in numerical terms. In this chapter we will examine the concept of **temperature** and the origins of **thermometry**—the art of measuring hotness and expressing it in numbers.

HOTNESS BY HAND

Let's recapitulate what we have thus far noted about mankind's adventures with heat through the sixteenth century. By his sense of touch Charlie could distinguish between hot and cold. He knew that when a hot object and a cold object were brought together they interacted—the hot object cooled off and the cold object warmed up. He was aware that some materials could inhibit or slow down that interaction. He discovered that if he performed mechanical work in the right way, for example, by rubbing two dry sticks together, he could heat an object without bringing it into contact with a still hotter object. He learned that heat interactions could bring about a wide range of

Call St. George? A real fire breather?
Just a dinosaur with fever!

The water nearly numbs his digit.
For ablutions—much too frigid.

physical and chemical changes. By rudimentary means Charlie had unwittingly even extracted mechanical work from a heat source (the sun). His descendants had substantially expanded the harnessing of heat to perform work. In qualitative terms, by 1700 mankind's perceptions and practices included every one of the essential heat phenomena. In an operational sense the only application that has since been added is the use of work for cooling as well as heating. (The mechanical refrigerator came much later than the heat engine. It was not until the middle of the eighteenth century that William Cullen, a Scottish physician, found he could make ice by pumping away the vapor from water in a closed container.)

All of this knowledge was purely empirical and qualitative. It permitted only the crudest kinds of predictions. There was acquaintance with heat phenomena without real understanding. No quantitative relations had been introduced. As Lord Kelvin was to say, anything that cannot be expressed in numbers we don't know very much about. The underlying quantitative concept in the heat game is temperature. We have been very careful thus far to avoid any reference to that word because it implies a measure of what we have called hotness, and no such measure was available during early human experience with heat phenomena. Charlie had no thermometers. He was almost entirely dependent upon his sense of touch to qualify hotness or coldness.

Although touch is quite a sensitive means of distinguishing between hot and cold, it can really only detect *differences* in **temperature,** the term we will use from now on to denote the degree of hotness of an object. The tactile sense provides no numerical value. Moreover, it is highly subjective and easily confounded. Immediately preceding experience can prejudice its response. If you have been throwing snowballs without gloves on, water from the cold tap can seem scalding. Texture can also introduce a bias. A sheet feels cooler than a blanket at the same temperature. In sum, a finger is not a very objective or quantitative thermometer. It has a very limited range. It cannot tolerate temperatures even as high as boiling water and becomes numb at temperatures near the freezing point of water.

We should note that the tactile range can be extended in a qualitative way by the sense of sight. Surely it was known early on that the degree of an object's hotness was indicated by its color. The terms *white hot* and *dull red heat* have been in the language for a long time (as has the expression *blue with cold!*). In addition it must also have been noted that temperature was reflected in the melting of various materials. Lead didn't liquefy until it was much hotter than the melting point of ice. Gold required still higher temperatures for fusion. But all of these observations couldn't provide a number, and a number is what we need.

EARLY THERMOMETERS

The first recorded step in the development of a numerical temperature scale was embodied in Galileo's thermoscope, shown schematically in Figure 2-1. If the bulb at the top came into contact with something warm, the air would expand and force the liquid downward. Conversely, if the bulb were cooled, the air would contract, and the liquid would rise in the tube. Thus, changes in the height of the liquid column indicated changes in the temperature of the bulb. As we will soon learn, expansion and contraction of a gas as the temperature changes ultimately became the basis for our present standard scale of temperature. Even so, Galileo's device left much to be desired. Although he did not know about barometers, you and I understand that the height of the liquid in the tube depends upon the pressure of the atmosphere. Consequently, at the same temperature the reading at the top of a mountain would be different from the reading at the bottom. Daily variations in barometric pressure would render it erratic. In sum, Galileo's invention could *indicate* temperature and temperature changes, but it was not very good for measuring them. For this reason, it is better called a thermo*scope* than a thermo*meter*.

The image that the word *thermometer* conjures up in almost everybody's mind is of the indexed, glass capillary tube with a bulb containing liquid at one end. With such a thermometer your mother took your temperature when you were sick. One hangs on the wall of almost every living room. No

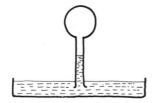

Figure 2-1   Galileo's thermoscope. It could indicate temperature changes but could not measure them very well.

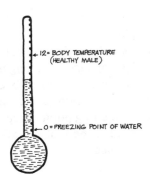

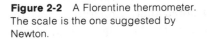

Figure 2-2 A Florentine thermometer. The scale is the one suggested by Newton.

chemist's laboratory is without several. These ubiquitous liquid-in-glass devices were developed in the middle of the seventeenth century at the Academy of Science in Florence and thus were long known as Florentine thermometers. Somewhat larger than today's models, they comprised a bulb filled with liquid, usually alcohol, attached to a glass tube of uniform bore whose length was marked off in equal intervals (Figure 2-2). As originally conceived, the volume between any two marks on the tube was supposed to be 1/1000 of the volume of the bulb. This specification put a substantial strain on the glassblower's art, and it soon became standard practice to apply the scale after the thermometer had been assembled and filled. The stem was marked at the position of the liquid level when the bulb was at some reference temperature, then again at a second reference temperature. The length of stem between the two reference marks was then divided by additional marks into equal intervals, or "degrees."

The choice of reference temperatures has an interesting history. In 1701 Isaac Newton proposed that the freezing point of water be taken as the lower reference point and given the value zero. The body temperature of a "healthy male" would be the higher reference point and given the value of 12. A few years later, Gabriel David Fahrenheit, the German instrument maker who first used mercury as a thermometric liquid, proposed that the zero point be taken as the lowest temperature that could be reached with a mixture of ice and salt. The human body temperature would still be 12. Such large "degrees" were somewhat awkward, so they were divided into eighths to give 96 degrees between the temperature of the ice–salt mixture and the normal body temperature.

Fahrenheit had shown that when ice and water coexisted, the temperature was constant and reproducible. He also demonstrated that the temperature at which water boiled was always the same at the same barometric pressure. On one of his thermometers, which had a scale linearly extrapolated beyond the body temperature point, the boiling point of water occurred at the 212-

degree mark and the freezing point at the 32-degree mark. Not long after his death, these two fixed points were generally adopted as standards. The Fahrenheit scale, based on 32 as the freezing point and 212 as the boiling point of water, became widely accepted. It is still the most commonly used one in English-speaking countries. You may wonder that the body temperature on this scale is today 98.6° instead of 96°. Normal body temperature has not changed. What is a fever for us would probably have been a fever for Charlie, but we can't do the experiment to prove it. The discrepancy probably stems from a lack of uniformity in the bore of Fahrenheit's original thermometer tubing. (But he might have had a chill on the day he was calibrating!)

The French got into the act when their great naturalist, R. A. F. de Réaumur, proposed dividing the interval between the freezing and boiling points into 80 degrees. His scale is still in use in some parts of Europe, but it is *not* the scale identified by the letter *R*. When it comes to temperature scales, *R* is short for Rankine, the Scottish engineer in whose honor the absolute Fahrenheit scale is named, about which more later.

Division of the interval between the freezing and boiling points of water into 100 degrees to form a centigrade scale was proposed by several scientists but has become associated with Anders Celsius, the Swedish astronomer. Since 1954, by action of the Tenth International Conference on Weights and Measures, the scale reading 0 at the freezing point of water and 100 at the boiling point is officially the *Celsius scale*. The symbol °C now means "degrees Celsius"; it is no longer an abbreviation for "degrees centigrade." Just why this honor was accorded Celsius is a bit puzzling. On the scale he proposed, the boiling point of water was to be 0 and the freezing point was to be 100!

BUT WHAT IS TEMPERATURE?

The development of liquid-in-glass thermometers and the specification of temperature scales provided a well-defined operation for the measurement of the quantity we call temperature. It would seem, therefore, that the concept of temperature as a fundamental property had thus been well established. Things are often not as simple as they seem, however. In the work of these early investigators, and in our discussion of it, there has been an implicit assumption for which there is no a priori justification. If you stick a thermometer in a bucket of water and find its temperature is 39°C and then take the temperature of a brick and find it is also 39°C, you expect that if you drop the brick in the water, neither temperature will change. You expect no change because in your experience when objects of the same hotness are brought together, no change in hotness occurs. But this absence of change reflects a particular characteristic of the real world that is the basis for the concept of temperature. You are *assuming* that the brick, the water, and the thermometer have something in common because the thermometer reads 39°C when in contact with either object.

This realization, the cornerstone of thermometry, came long after the formulation of the *First* and *Second Laws of Thermodynamics*. R. H. Fowler, in 1931, was the first to point it out. The formal statement of this fundamental assertion is called the **Zeroth Law of Thermodynamics** in honor of its basic role. It says that if object *A* does not change its hotness when brought into contact with object *B* and if object *C* does not change its hotness when brought into contact with object *B*, then *A* and *C* will not change their hotness when brought into contact with each other. Or more tersely:

Two objects that are in thermal equilibrium with a third object will be in thermal equilibrium with each other.

These statements may appear trivial to you, but they seem so only because they are so consistent with your experience with hotness. In his innocence even Charlie would have taken this proposition for granted. But, in general you certainly cannot expect that a relation applying between *A* and *B* and between *A* and *C* will inevitably apply between *B* and *C*. Two boys who like the same girl will not necessarily like each other. Water is miscible with alcohol and so is gasoline. But gasoline and water won't mix. Thus, the Zeroth Law is a statement about the nature of the world not derivable from a more primitive or elementary proposition. It is a genuine *law* that provides the basis for much of our subsequent discourse. Meanwhile, there is another less abstract but more nagging difficulty with the concept of temperature and the operations to measure it that we have thus far described.

Suppose you have two liquid-in-glass thermometers, one filled with alcohol, the other with mercury. They have both been given Celsius scales by putting marks at the liquid level when they are in ice water and again when they are in boiling water. The interval between the two marks is graduated in 100 equal degrees. Now suppose you stick the mercury thermometer in a bucket of water and find that it reads 50°C. If at the same time you also put the alcohol thermometer in the bucket, you would find that *it* reads about 48.5°C!

Even more exasperating would be the reading given by a thermometer in which water was the liquid. (Of course, you would only be able to use it at temperatures above 0°C.) If the bucket of water were at 4°C as indicated by the mercury thermometer, the water thermometer would show a "temperature'' of less than zero (because water happens to have its maximum density and, therefore, its minimum volume at 4°C). Which thermometer is right? At this point in our discourse, the answer would have to be whichever one we choose to be right. That is, to measure temperature with such a device we must specify the thermometric liquid. We can decree that the mercury thermometer will be the standard, or we can choose another liquid. Whatever liquid we use, *all* thermometers would agree at the fixed points, of course. But all other temperature readings from thermometers filled with other than the standard liquid would have to be corrected. If we are dependent for the measurement operation upon behavior of a particular substance, chosen arbitrarily, we would probably be a bit uneasy about the nature of temperature itself. After all, if value of temperature depends upon the substance we choose to measure it with, is there really anything absolute, fundamental, and universal about it as a property?

THERMOMETRY WITH GASES

To extract ourselves from this dilemma, let's now consider the use of gas as a thermometric substance. Clearly, we can't use it in quite the same way that we use liquids. Gas entirely fills any vessel that confines it. It displays no free surface or interface. Its volume is the volume of its container. When its hotness is increased, however, a gas will expand, or increase its volume, if the container has elastic walls so that the pressure can remain constant. Conversely, if its volume is kept at a constant value, pressure will increase as hotness increases. These empirical observations of the French physicists J. A. C. Charles (1787) and J. L. Gay-Lussac (1802) became the basis for laws that we will discuss in the next chapter. For the moment, we will simply assert that the pressure of a gas at constant volume increases with rising temperature.

In the apparatus of Figure 2-3, a line engraved on the glass tube (at the arrow) defines a given volume of gas whose pressure will vary as the temperature of the surrounding bath varies. The pressure associated with the given volume at the various temperatures, i.e., required to maintain the **meniscus** (gas–liquid interface) at the engraved line, is the observed thermometric property. It is measured by the height of the fluid in a manometer, the U-shaped tube filled with liquid in the sketch. (For more details on measuring pressure with manometers see Appendix I.) It is to be understood that the diagram is only schematic. A real gas thermometer is extremely intricate and complicated to use. Allowances must be made for variations in the volume of the containing bulb as the temperature changes, for the contribution to total pressure of the vapor pressure of the liquid used to define the

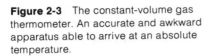
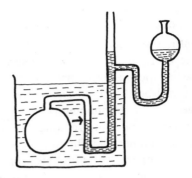

Figure 2-3 The constant-volume gas thermometer. An accurate and awkward apparatus able to arrive at an absolute temperature.

volume, for the variation with temperature of the liquid density, and so on. Nevertheless, the principle is straightforward even if the practice is complicated.

Clearly the pressure indicated by the manometer will be higher when the bath comprises boiling water, (is at the *steam point*) than when the bath contains coexisting ice and water (is at the *ice point*). It should also be clear that we can arbitrarily define a temperature ratio in terms of a pressure ratio by writing:

$$\frac{T_s}{T_i} = \frac{p_s}{p_i} \tag{1}$$

where the subscripts indicate the steam point and the ice point. If we try out this relationship with a number of different gases, say helium, nitrogen, argon, and methane, starting in each case with a pressure of about 1 atmosphere at the ice point, that is, $p = 760$ torr when $T = T_i$, we get nearly the same value for p_s/p_i, no matter which gas is in the thermometer. This consistency is reassuring in that there is almost no dependence of the temperature ratio determination upon the particular choice of thermometric substance— at least among these several gases.

Let us now recognize that we can also vary the amount of gas in the bulb so that the pressure at the ice point, p_i, can be any value we choose. We will find that the ratio of pressures at the steam point and the ice point, p_s/p_i, will depend to some extent upon the amount of gas present, that is, upon the pressure at the ice point, p_i. If we fool around enough, we will discover what a number of careful investigators have found—that as the initial pressure, p_i, gets smaller and smaller, the value of the ratio, p_s/p_i, for various gases gets closer and closer to the same value. If we plot the value of this ratio against the value of p_i (determined by the amount of gas in the bulb) for several gases, we get the results shown in Figure 2-4. As p_i approaches zero, as we extrapolate to the vertical axis, *all* the gases give *exactly* the same limiting

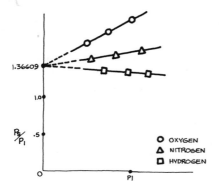

Figure 2-4 Measurements obtained with a constant-volume gas thermometer. In the limit of very low pressure (density), all gases give the same extrapolated value for the ratio p_s/p_i.

○ OXYGEN
△ NITROGEN
□ HYDROGEN

value for p_s/p_i, namely, 1.36609 ± 0.00004. This observation, which has been confirmed for every gas tried, means that we get the same value for the ratio of two temperatures no matter what the chemical composition of the gas. With confidence, we can now define a temperature scale by asserting that for the two temperatures T_s and T_i:

$$\frac{T_s}{T_i} = \lim_{p_r \to 0} \frac{p_s}{p_i} = 1.36609 \tag{2}$$

This equation is not a complete definition of the temperature scale because there are two free quantities, T_s and T_i, but only one relation. Accordingly, we also assert that:

$$T_s - T_i = 100 \tag{3}$$

This last condition makes the *size* of the degree the same as for the Celsius scale, which provides that $T_i = 0$ and $T_s = 100$. If you solve Equations (2) and (3) simultaneously you will find with simple algebra that $T_s = 373.16$ and that $T_i = 273.16$.

For any other temperature T, corresponding to pressure p, we can write:

$$\frac{T}{T_i} = \lim_{p_r \to 0} \frac{p}{p_i}$$

or

$$\tag{4}$$

$$T = T_i \lim_{p_r \to 0} \frac{p}{p_i} = 273.16 \lim_{p_r \to 0} \frac{p}{p_i}$$

In other words, to find the temperature of an object with a gas thermometer scale, we find the pressure p exerted by a gas at a given volume when it has been near the object long enough for thermal equilibrium to occur (practically, this means when the pressure no longer changes with time). We also find the pressure p_i exerted by the same amount of gas at the same volume when it is in thermal equilibrium with a mixture of ice and water. To obtain the temperature T of the object we multiply the ratio of the two pressures p/p_i by 273.16. To be precise, we must take the limiting value of the ratio as we decrease the amount of gas in the given volume.

THE INTERNATIONAL SCALE

Obviously, the gas thermometer is not a very convenient device. It is bulky, awkward, intricate, and tedious to use. A single, reliable, temperature measurement using gas thermometry may take months. Consequently, it is an art practiced in only a few laboratories in the world (the National Bureau of Standards in Washington, D.C., is one). For everyday use the Seventh General Congress on Weights and Measures in 1927 adopted a much more convenient reference scale. Revised several times since, it is meant to be as close as possible to the gas thermometer scale and makes use of a number of fixed points that have been carefully measured in laboratories skilled in the art of gas thermometry. The most recent revision of this International Practical Temperature Scale was agreed upon by the International Committee of Weights and Measures in October 1968. Referred to as the IPTS-68 and identified by the symbols t_{68} (for the Celsius scale) and T_{68} (for the Kelvin scale, which we will describe shortly), it is defined by the fixed points set forth in Table 2-1.

In addition to these fixed points, the International Scale provides well-defined interpolation procedures for obtaining intermediate temperatures. These procedures make use of platinum resistance thermometers and platinum—rhodium thermocouples. The resistance thermometer takes advantage of the fact that the electrical resistance of a pure metal increases in a very reproducible way as temperature increases. Thermocouples comprise two wires of dissimilar metals, for example, platinum and an alloy of platinum and rhodium, that are fused together at each end. One of the wires is then cut in the middle, and each of the resulting ends is connected to the terminals of a device that measures electromotive force, that is, a potential difference. The measured emf, or voltage, is small but is very reproducibly dependent upon the temperature difference between the two junctions or fused ends. Because it is easy to make electrical measurements of resistance

TABLE 2-1 Fixed points for the International Practical Temperature Scale, 1968.

Equilibrium state	t_{68}K	t_{68}°C
Triple point of equilibrium hydrogen*	13.81	−259.34
Boiling point of hydrogen at 25/76 atm	17.042	−256.108
Boiling point of hydrogen at 1 atm	20.28	−252.87
Boiling point of neon at 1 atm	27.402	−246.048
Triple point of oxygen	54.361	−218.789
Boiling point of oxygen at 1 atm	90.188	−182.962
Triple point of water	273.16	0.01
Boiling point of water at 1 atm	373.15	100
Freezing point of zinc	692.73	419.58
Freezing point of silver	1235.08	961.93
Freezing point of gold	1337.58	1064.43

* Triple point is temperature at which solid, liquid, and vapor coexist.

and potential with great accuracy, resistance thermometers and thermocouples are widely used as temperature sensors.

In addition to the primary reference points given in the table, the International Scale specifies a number of secondary reference temperatures based on melting points and boiling points of various substances. These well-defined primary and secondary reference temperatures together with the prescribed means for obtaining intermediate temperatures make it relatively easy for any laboratory to calibrate its thermometers and to determine temperatures in terms of the International Scale with negligibly small errors.

What we have been calling the ice point, T_i, does not appear in today's definition of the gas thermometer scale and therefore is not in Table 2-1. In its place is the so-called **triple point** of water, T_{tp}, the temperature at which ice, liquid water, and water vapor coexist. The **ice point,** defined as the temperature of a well-stirred mixture of ice and water saturated with air at 1 atmosphere pressure, had been difficult to reproduce. Air has a small but nonnegligible solubility in water. (Any substance dissolved in water will lower its freezing point.) But the water from melting ice has no air in it. Consequently, a stirred mixture of air-saturated water and melting ice is not really in a well-defined equilibrium state because the actual state depends upon the rate of stirring, that is, how much air is dissolved.

After years of discussion, the General Conference on Weights and Measures in 1954 finally adopted a suggestion originally proposed by James Joule and William Thomson (Lord Kelvin) 100 years earlier. This suggestion was to use a single point as the basis for completing the definition of the gas thermometer scale defined by Equation (2), that is, to replace Equation (3) by a single reference temperature. The triple point of water was chosen as this single, fixed point. It is the temperature at which ice, water vapor, and liquid water coexist in the absence of air or any other substance that is appreciably soluble in water. Its value was taken as 273.16 degrees. This triple point temperature is easily reproducible to within $\pm 0.00008°C$ and provides a very convenient absolute standard. Consequently, the gas thermometer scale is now defined by the following pair of equations:

$$\frac{T}{T_{tp}} = \lim_{p_{tp} \to 0} \frac{p}{p_{tp}} \tag{5}$$

$$T_{tp} = 273.16 \tag{6}$$

Interestingly enough, on this scale the ice point has the temperature 273.15, only 0.01 degrees different from the value we obtained using Equations (2) and (3). You might think this is all much ado about nothing. Charlie certainly would. To be sure, at the melting point of ice an uncertainty of 0.01 degrees is only about 0.003 percent. But at very low temperatures, where, for example, research on superconductivity in metals and superfluidity in liquid helium is carried out, the percentage error can become substantial. Today many measurements are made at temperatures as low as 0.01 degrees. There an uncertainty of 0.01 is 100 percent! (Charlie remains skeptical!)

We will learn later that the gas thermometer scale is identical with what is known as the *absolute thermodynamic scale*, which is entirely independent of any thermometric substance. This absolute scale is called the Kelvin scale in honor of William Thomson, the Scotsman who became Lord Kelvin and who did much to establish modern thermodynamics. Consequently, we will use upper case T, the customary symbol for absolute temperature, and indicate specific values as numbers followed by the letter K, for example, 98 K. You should understand, however, that our use of this symbolism presumes an identity we have not shown with a scale we have not yet defined!

When temperature is elevated Aspirin is indicated.

The equivalent absolute Fahrenheit scale is known as the Rankine scale in honor of the Scottish engineer, William John MacQuorn Rankine, a contemporary of Kelvin and one of the founders of modern thermodynamics. Temperatures on this scale are identified by the letter R. For example, 491.69 R = 273.16 K. We can define it using Equation (2) and an equation specifying the Fahrenheit degree size:

$$\frac{T_s}{T_i} = \lim_{p_r \to 0} \frac{p_s}{p_i} = 1.36609 \tag{2}$$

and

$$T_s - T_i = 180 \tag{7}$$

Here, we find T_i = 491.68 and T_s = 671.68. Since 1968, however, the Fahrenheit and Rankine scales have been defined in terms of the Kelvin scale by asserting that 1 R is precisely 5 K/9. Thus, at the ice point, T_i = 491.67 R and t_i = 32.00°F. At the triple point of water, T_{tp} = 491.69 R and t_{tp} = 32.02°F.

SUMMING UP

The important message of this chapter has been cloaked in a fair amount of historical garb. If these anecdotal trappings are removed, we are left with the following essential features of temperature and its measurement:

1. *Temperature* is a quantitative measure of the hotness of an object. Its existence is a characteristic of the real world that provides the basis for the *Zeroth Law of Thermodynamics:* If two objects do not change their hotness when brought into contact with a third object, they will not change their hotness when brought into contact with each other.

2. *Thermometers* are devices that, when brought into contact with an object, indicate its temperature in terms of some observable property. The length of a liquid column, the pressure of a gas at constant volume, the resistance of a wire, and the electrical potential difference between two junctions of different metals, are frequently used *thermometric properties*.

3. A *temperature scale* is established by choosing two suitable reference temperatures and assigning a value to each or assigning a value to one and asserting a value for either the ratio of the two or the difference between them. The freezing and boiling points of water (at 1 atmosphere pressure) have often been used as reference temperatures. On the popular Fahrenheit scale, the freezing point of water is 32°F and the boiling point is 212°F. On the more widely used Celsius scale, the freezing point is 0°C and the boiling point is 100°C.

4. The *Ideal Gas Thermometer Scale* has become an almost universally accepted standard scale. It is defined by the relation:

$$\frac{T}{T_{tp}} = \lim_{p_{tp} \to 0} \frac{p}{p_{tp}}$$

where p is the pressure of a given volume of gas at a temperature T, and T_{tp} is the temperature at the triple point of water (where ice, liquid water, and water vapor coexist). T_{tp} is assigned the value of 273.16. Precise work requires taking the limiting value of the ratio p/p_{tp} as the amount of gas in the constant volume becomes vanishingly small.

5. The Gas Thermometer Scale is an absolute scale because its definition provides for a true zero. Because it is identical with the absolute *Kelvin thermodynamic temperature scale,* which is independent of the properties of any substance, we will use K to indicate absolute temperatures obtained by gas thermometry. The absolute scale based on a degree of the same size as the Fahrenheit degree is defined in terms of the Kelvin scale by the relation: 1 R = 5 K/9 (exactly). Thus, *Rankine* temperatures can be obtained by adding 459.67 to the Fahrenheit reading. Kelvin temperatures can be obtained by adding 273.15 to the Celsius value.

6. Because gas thermometry is a difficult art, we have, by international agreement, an *International Temperature Scale.* It is meant to be as nearly identical as possible with the gas thermometer scale but is defined by assigning values obtained by gas thermometry to easily reproducible points over a wide range of temperatures. These points are usually characterized by the boiling, triple, or melting points of various pure substances. In addition, the definition of the scale provides interpolation procedures between the fixed points. Thermocouples and platinum resistance thermometers are used in the interpolation procedure.

SCALE PRACTICE

1. List as many observable quantities as you can think of that vary with temperature and that might be used to measure it. Don't be inhibited by practical considerations. For example, the pitch of a guitar string depends upon its tension and, therefore, its temperature, but would be an awkward basis for a thermometer.

2. The normal (1 atm) boiling point for helium is the lowest of any substance at $-268.93°C$. Tungsten has the highest melting point of any metal at $3410°C$. What are the corresponding temperatures on the Kelvin, Fahrenheit, and Rankine scales?

3. At what Kelvin temperature do Celsius and Fahrenheit temperatures have the same numerical value?

4. In a liquid-in-glass thermometer, the length of the liquid column is 80 mm when the thermometer is dipped in boiling water. It is 40 mm long when the thermometer is in ice water at $0°C$. What is the length of the column when the thermometer is immersed in a liquid at $40°C$? 200 K?   110°F?   620 R?

5. In determining the melting point of a metal alloy with a gas thermometer, an investigator finds the following values for the gas pressure p when the pressure p_{tp} at the triple point has the indicated value. (Pressures are in torr. One atm equals 760 torr.)

p_{tp}	100.0	200.0	300.0	400.0
p	233.4	471.6	714.7	962.9

Find the limiting value of the ratio p/p_{tp} as p_{tp} approaches zero. If the triple point temperature is 273.16 K, what is the melting temperature of the alloy?

6. If an ideal gas is used in a constant volume gas thermometer, $T_s/T_{tp} = p_s/p_{tp} = 1.36605$ and in general $T_1/T_2 = p_1/p_2$. Subscripts s and tp refer to steam and triple points, 1 and 2 to any two temperatures. Find T_s and T_{tp} on an absolute "Newtonian" scale for which $t_s - t_{tp} = 23$. What is the Newtonian temperature at absolute zero if $t_{tp} = 0°$ on the nonabsolute Newtonian scale?

7. If the centigrade scale that Anders Celsius originally proposed had been adopted, the temperature at the ice point would be $100°C$ and the

temperature of the steam point would be 0°C. The gas thermometer scale might then have been defined by the following two equations:

$$\frac{T_i}{T_s} = \lim_{p_i \to 0} \frac{p_s}{p_i} \quad \text{and} \quad T_i - T_s = 100$$

and in general $T_i/T = \lim_{p \to 0} p/p_i$. On such a scale what would be:

(a) the temperature of the ice point?
(b) the temperature of the steam point?
(c) the temperature corresponding to absolute zero Kelvin?
(d) the temperature corresponding to 10^8 K (about what will be required to achieve thermonuclear fusion)?

8. The electromotive force produced by a thermocouple can be described by the equation:

$$\mathscr{E} = a + b(t - t_o) + c(t - t_o)^2$$

where a, b, and c are constants for a particular thermocouple, t is the Celsius temperature, and the subscript o relates to the ice point, 0°C, or 273.15 K. $\mathscr{E}$ is the electromotive force (emf).

(a) In terms of a, b, and c define a centigrade scale (100 degrees between ice and steam points) that is linear in $\mathscr{E}$.
(b) Find an expression for $\mathscr{E}$ in terms of a, b, c, and T where T is the Kelvin temperature. What is $\mathscr{E}$ at absolute zero?
(c) If we define a θ temperature scale by $\theta/\theta_i = \mathscr{E}/\mathscr{E}_i$ where $\theta_i = 200$, what is the value of θ at $t = -100°C$?

9. Suppose we define a temperature scale by the equations

$$\frac{T_s}{T_i} = \lim_{p \to 0} \frac{p_s}{p_i} = 1.36609$$

as in gas thermometry, but that we take the Réaumur scale for the additional relation needed to specify T_s and T_i, the steam and ice points, that is, $T_s - T_i = 80$.

(a) Find the numerical values on the absolute Réaumur scale for the steam point and the ice point.
(b) If the ice point on the conventional Réaumur scale is $t_i = 0$, what will t be at absolute zero?

(c) Lead melts at 327°. What is its melting point on the absolute
Réaumur scale?

10. Define a temperature scale as in the preceding problem except that in
place of the Réaumur scale for the second relation assert that $T_s = 100$.
Find the temperature on this scale for the ice point, the steam point,
and the melting point of lead (327°C).

Might anglers not tell such tall tales
If they used calibrated scales?

3 Systems, Properties, and States

In the last chapter we traced the development of thermometry, the art and science of assigning a numerical value to the hotness of an object. Temperature is the name given to this quantitative measure. Now we will examine the relationship between temperature and other characteristics of objects or systems that can be numerically expressed and that we call *properties*. This examination will lead us to the idea of an *equation of state,* an algebraic relation between property values.

SPECIFIC MEANINGS FOR SOME GENERAL TERMS

In one sense, any branch of science can be regarded simply as a collection of definitions, together with the relations among them, that describe the systems with which that particular branch of science is concerned. The word **system** stems from the Greek words that mean to bring together, or combine. It has many contextual meanings. In our discourse, the term *system* will refer to that part of the real world on which we focus our attention. It may be a container of gas, a bucket of water, a cloud, an engine, a plant, or a planet. Frequently, we define a system by its **boundary,** a sort of a magic cloak to which we can ascribe almost any desired characteristic. It may be absolutely rigid or completely elastic; impervious to matter or permeable; a perfect thermal insulator or an ideal conductor. It is immaterial, contributing neither mass nor substance to the system or the surroundings. It is often convenient to identify the boundary with a closed geometric surface on one side of which is the system, on the other the rest of the real world, i.e., the **surroundings**. On plane figures we often represent such a boundary surface by a simple closed curve of random shape.

Consider **mechanics,** the branch of science that concerns itself with systems whose complete quantitative description can be reduced to relations

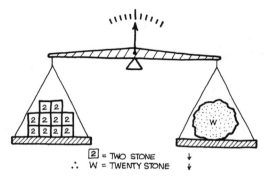

Figure 3-1 An example of property measurement. The mass of a subject system is determined by comparing its weight with the weight of a reference mass.

2 = TWO STONE
∴ W = TWENTY STONE

between mass, length, and time. Force is sometimes regarded as an independent quantity, but if we accept Newton's laws of motion, we can also express it in terms of mass, length, and time. (See Appendix I for definitions and units.) These quantities, and various combinations of them, such as area (length × length), pressure (force/area), volume (length × length × length), and density (mass/volume) are all numbers obtained by prescribed and well-defined operations. For example, the **length** of an object is the number of times we can lay a standard stick in adjacent positions from one end of the object to the other. Of course, we usually use a graduated tape or ruler to do the counting for us. Likewise, the **weight** of an object can be measured by finding how many standard weights it will offset on a balance, or, more easily, how far it will stretch a spring from which it is suspended and whose stretchability has been determined with standard weights (Figure 3-1). The quantities identified by these prescribed measuring operations are called **properties.** The **state** of a system is defined by the numerical values of its properties.

It seems likely that Charlie engaged in this numbers mensuration game to some extent as soon as he learned to count. He probably reckoned time in days (suns), months (moons), and years (seasons). Smaller time intervals had to be gauged in analogous terms by the position of the sun during the day and the moon or stars at night. Distances could be measured by counting stick lengths, steps, or days of travel. He could measure volumes of water by

German, Japanese, or Swiss,
A watch is handier than this.

Till this critter weighs in fatter
It's not ready for the platter.

shellfuls and seeds by handfuls. Mass or weight gave him more trouble until he grasped the concept of balance and contrived some scales.

In the previous chapter, we learned that temperature, like length, is also a property number resulting from a prescribed set of operations. As in the case of length, which can be expressed in a variety of different units (arbitrarily defined standard unit lengths such as inches, meters, rods, or light years), temperature can also be expressed in terms of any of several arbitrary scales. But temperature is different in kind from mechanical properties. It cannot be reduced to, or completely expressed in, terms of mass, length, and time. Even though we may measure a system's temperature by a mechanical property, such as the length of a liquid column or the pressure of a gas, the property *temperature* embodies more than that measured mechanical variable. To complete a description of temperature from the measurement of this variable we always need a reference value of that variable at some particular, reproducible hotness, associated, say, with the melting of ice or the boiling of water. Thus, temperature adds a dimension additional to mass, length, and time for the description, treatment, and analysis of systems.

WHY TEMPERATURE GETS INTO THE ACT

Let's examine the case of an *ideal* simple pendulum. Its *motion* and *period* are determined entirely by the length of its suspension and the strength of the gravitational field. The temperature of the bob is of no importance and does not appear in the equations describing its behavior. But we all know that the motion of a *real* pendulum is only approximately described by the ideal law (see Figure 3-2). A real pendulum gradually comes to a stop even if it is in a vacuum where there is negligible air resistance. The length of suspension and the magnitude of the gravitational field do not change during the stopping process. Somehow, the treatment, or description, in purely mechanical terms is incomplete. What happens, of course, is that what we call "friction

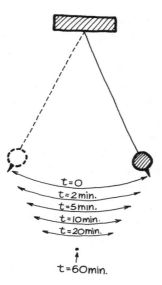

Figure 3-2 Real pendulum behavior: The amplitude of the swing will gradually decrease to zero. A truly complete description of this process involves temperature as well as the purely mechanical properties sufficient to describe an ideal pendulum (which never stops!).

in the suspension'' makes the system dissipative, or nonconservative. According to mechanics, the motion analysis of an ideal pendulum applies strictly to a conservative system, that is, one without friction or any other kind of dissipation.

Actually, our use of the words *friction* and *dissipation* has been a bit glib. Many of us have an intuitive feeling for these terms, but we would be hard put to explain them very precisely. For the moment, we will simply recognize that friction and dissipation are phenomena that, in the absence of a continuous work input, damp out motion in real systems—perhaps slowly, but always, inexorably, and completely. More importantly, for our present purpose, there are always other observable changes associated with this slowing down.

If we carefully observe a real pendulum, we will note that as it slows down, the suspension warms up. Similar temperature rises occur when bouncing balls, spinning wheels, and other kinds of real moving systems come to rest. Of course, temperature rises also occur in these systems when their motion is maintained by a continuous work input. In such cases, the temperature usually climbs until it reaches some steady state value at which the rate of heat interaction with the surroundings equals the rate at which heat is generated by the work input. It would seem, therefore, that a *complete* analysis of real systems must take into account any temperature change or heat interaction that occurs.

Indeed, temperature makes *possible* a comprehensive treatment of a mechanical system even when it is nonconservative. Including temperature changes and associated heat interactions in the accounting also allows a more complete analysis of other kinds of systems and phenomena in which dissipation occurs. When currents flow in electrical circuits, resistance causes temperature changes. In fluid flow, viscosity results in temperature increases. These temperature and heat effects are usually incidental nuisances in the disciplines of electromagnetism and mechanics. They are attributed to the "imperfections" of real systems. But they are vital features of **thermodynamics,** the subject whose main preoccupation is with temperature differences and their relation to heat and work interactions. Its name compounded from the Greek words for heat and power, thermodynamics is what this book is really all about. With roots in a concern about the efficiency of heat engines, that is, how much fuel will produce how much work, it has evolved into a generalized method of accounting, or bookkeeping, on all kinds of systems especially when heat interactions are involved and temperature is an important variable.

RELATIONS WITH TEMPERATURE

We have pointed out that temperature is a property, or quantitative characteristic, of a physical system. It is in addition to, and fundamentally different from, other properties that can be expressed in terms of mass, length, and time. But we have also seen that the very operations by which we measure temperature stem from the dependence of other properties upon it. The liquid-in-glass thermometer works because liquids expand when they are heated. Other, less familiar thermometers take advantage of changes in electrical properties that occur when the temperature changes. The thermocouple makes use of the electromotive force, that is, the voltage difference, generated when two bimetallic junctions are at different temperatures. The gas thermometer, which provides the most reliable temperature measuring because it gives the same value no matter what gas is used, is based on the proportionality between temperature and pressure in a fixed volume of gas.

Let's examine further the dependence of some properties upon temperature by recalling the equations that define the gas thermometer scale:

$$\frac{T}{T_{tp}} = \lim_{p_{tp} \to 0} \frac{p}{p_{tp}} \quad \text{and} \quad T_{tp} = 273.16$$

We confine our attention for the moment to cases in which the pressure is low enough so that $T/T_{tp} = p/p_{tp}$, which we can rewrite as:

$$p = (p_{tp}/T_{tp})T$$
$$= KT \tag{1}$$

where K is a constant equal to the ratio of p to T at the triple point (where ice, liquid water, and water vapor coexist). This relation, which holds for a *constant volume* and for a fixed amount of gas in the thermometer, embodies two empirical laws that were enunciated long before gas thermometry became an established art.

The first law was announced by the English natural philosopher, Robert Boyle, in the early 1660s. **Boyle's Law** says that the pressure of a gas at constant temperature is inversely proportional to its volume. If the pressure is doubled, the volume is halved, and vice versa. A plot of pressure against volume at constant temperature, as shown in Figure 3-3, has the form that mathematicians call a *hyperbola*. We will usually refer to it as an **isotherm** in recognition of its role as the locus on a pV diagram (for a fixed amount of gas) of all points with the same temperature. Whatever its name it is simply a graphic representation of the relation between p and V according to Boyle's Law. A decade or so later, this law was discovered independently by the French physicist, Edme Mariotte, and is sometimes called *Mariotte's Law*. (Toujours en France!)

The second of the empirical *gas laws,* as we noted earlier, was discovered in 1787 by Jacques Alexander César Charles, a French physicist who introduced the use of hydrogen in balloons. It was independently discovered and first published in 1802 by Charles' compatriot, Joseph Louis Gay-Lussac. **Charles'** (or Gay-Lussac's) **Law** asserts the linear dependence of gas volume

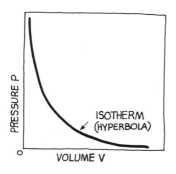

ISOTHERM
(HYPERBOLA)

PRESSURE P

VOLUME V

Figure 3-3   Boyle's Law. For a gas at constant temperature, the product pV is also constant. A plot of p against V thus takes the form of a hyperbola as shown here.

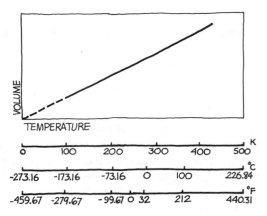

Figure 3-4 The law of Charles and Gay-Lussac. At constant pressure, the volume of a gas increases linearly with increasing temperature on any of the common scales. Only on an absolute scale, for example, Kelvin or Rankine, do V and T for an ideal gas become zero together.

upon temperature at constant pressure. Of course, these early experimentalists did not measure temperature in terms of an absolute scale. They used liquid-in-glass thermometers with empirical (e.g., Celsius or Fahrenheit) scales. It happens, as shown in Figure 3-4, that the volume of a gas at a sufficiently low constant pressure has a linear dependence on both of these empirical temperatures as well as on the absolute temperature. The difference is that on an absolute scale the zero value of volume (as obtained by extrapolation) occurs at the zero value of temperature. On the empirical scales, the zero value of volume occurs at negative temperatures because the temperature to which a zero value was assigned was arbitrary (e.g., the freezing temperature of water).

If we stick with absolute temperatures, we can write the two gas laws as follows:

$$\text{Boyle–Mariotte:} \qquad pV = \text{constant (at fixed } T)$$
$$\text{Charles–Gay-Lussac:} \qquad V/T = \text{constant (at fixed } p)$$

An expression including both of these relations is simply:

$$pV/T = \text{constant} \tag{2}$$

This general relation between pressure, volume, and temperature (absolute) holds for a fixed amount of any gas that obeys the laws of Boyle and Charles. Such a relation, linking two or more properties of substance, is called an

equation of state. Notice that Equation (2) reduces to Equation (1) when V is fixed.

MOLES AND MOLECULES

If you experiment a little with Equation (2), you will quickly verify what your intuition may have already told you: The value of the constant depends upon the amount of gas. At a given temperature and pressure, there must be twice as much gas in 2 cubic feet as there is in 1 cubic foot. Thus, in Equation (2) if we double V for a given p and T, the left-hand side will be doubled. Therefore, the value of the constant on the right-hand side must also be doubled if the equation is to remain an equation. Similarly, if we want to triple the pressure p at a fixed volume V and temperature T, we would have to triple the quantity of gas and, therefore, the size of the constant. To understand how the constant is evaluated, you must understand what is meant by the term *molecular weight*. If that concept is at all strange, read the accompanying sketch on the atomic theory of matter (see page 42). Otherwise, full speed ahead.

It is a matter of custom and convenience to relate the value of the constant in Equation (2) to that amount of gas we call a **mole** (abbreviated mol in dimensional formulas). A **mole** of any substance is that number of mass units (for example, grams, pounds, ounces, kilograms, or tons) equal to the molecular weight of the substance. Thus, 1 g mol of hydrogen is 2 g; 1 lb mol of water would be 18 lb; 1 ton mol of oxygen would be 32 tons; 1 oz mol of carbon dioxide would be 44 oz. If you think about it, you will realize that because the molecular weight scale is based on the relative weight of the molecules, 1 mole of any material must have the same number of molecules. This number is known as **Avogadro's Number.** In the case of 1 g mol, it has the value 6.023×10^{23}. Consequently, 1 kg mol, for example, 32 kg of oxygen, will contain 6.023×10^{26} molecules. These are incredibly large numbers. If each molecule in 2 g of hydrogen were turned into a grain of sand, there would be enough sand to cover the earth to a depth of 1 m!

Amadeo Avogadro was an Italian physicist who proposed in 1811 that equal volumes of gas under the same conditions of temperature and pressure contain the same number of molecules, no matter what the chemical identity of the gas. This hypothesis, sometimes called **Avogadro's Law,** has been verified for all substances under the conditions necessary for them to give the right temperature in a gas thermometer. In other words, it is true for all gases in the limit of very low pressures. It is very nearly true for many gases

THE ATOMIC THEORY: A THUMBNAIL SKETCH

Chemistry emerged from alchemy when chemists followed the lead of the great French scientist Antoine Lavoisier by making quantitative observations of chemical phenomena. In particular, they very carefully weighed the amounts of one substance that would react with another to form a new substance. As such data accumulated, the practical need to organize and correlate it in a useful and simple form became more pressing. There was also the urge to understand the great variety of observations in terms of some simple underlying principle. Early in the nineteenth century, the British chemist John Dalton wed quantitative analysis with the old Greek idea of atoms as the ultimate tiny and indivisible particles constituting all matter. His *Atomic Hypothesis* and the theory of matter to which it led so effectively described all manner of chemical and physical phenomena that scientists were persuaded of the reality of atoms long before there was direct evidence of their existence.

In the modern chemist's atomic theory all substances are made up of **atoms,** each with a core or nucleus comprising its **atomic number** of positively charged **protons,** sometimes with one or more neutrons having almost the same mass as protons but no charge. The nucleus is surrounded by its atomic number of satellite **electrons,** each with a negative charge and a mass only $\frac{1}{1840}$ that of a proton. An atom with a deficiency or an excess of electrons has net charge and is called an **ion.** The atomic number determines the chemical behavior of an atom and ranges in the naturally occurring **elements** from 1 for hydrogen to 92 for uranium. Thirteen "transuranium" elements with atomic numbers up to 105 have been made in the laboratory by bombarding nuclei of heavy atoms with neutrons. Actually, plutonium (atomic number 93) is "natural" because trace amounts have been found in some California rocks.

Atoms with nuclei of identical charge (and therefore chemical properties) but differing numbers of neutrons (and therefore mass) are known as **isotopes.** The atomic weight of an element is the mass of its atoms relative to the most abundant isotope of carbon, arbitrarily assigned the value 12. On this scale, for example, the atomic weight is 1 for hydrogen (H), 14 for nitrogen (N), 16 for oxygen (O), 63 for copper (Cu), 108 for silver (Ag), and 238 for uranium (U). (The letters in parentheses are the customary shorthand for these atoms.) A table of atomic weights is in Appendix IV. Its actual values are not integral, mostly because they are averages over the natural isotope abundance for each element. We will use the nearest integral value, for example, 108 for silver rather than 107.868.

Although there are only 93 elements, there are more than a million **compounds,** the name given to substances comprising more than one element. This great multiplicity occurs because atoms can form bonds with other atoms in almost countless combination, thus giving rise to stable clusters of atoms known as **molecules.** A molecule is the smallest individual particle of any substance that is stable and retains the characteristic features of that substance. Thus, a molecule of hydrogen contains 2 atoms and can be represented by the formula H_2. A molecule of carbon dioxide containing 2 atoms of oxygen and 1 of carbon has the formula CO_2. Octane, a major component of gasoline, has the formula C_8H_{18}, which means it contains 8 atoms of carbon and 18 of hydrogen. The **molecular weight** of a molecule is the sum of the atomic weights of its constituent atoms. Thus, the molecular weight of H_2, CO_2, and C_8H_{18} are respectively 2, 44, and 114.

at ordinary temperatures and pressures and accounts for a rule you may have encountered: 1 g mol of any gas at 0°C and 1 atm occupies a volume of 22,414 cm³.

THE IDEAL GAS LAW

Let's consider Equation (2) in light of Avogadro's Law. Because 1 mole of any species has the same number of molecules and because Avogadro's Law requires that the same number of molecules must occupy the same volume at a given temperature and pressure, it follows that the constant in Equation (2) must be the same for 1 mol of any gas. Indeed, this conclusion is verified by any number of experimental tests. Therefore, we accept it and write:

$$pV/T = R \qquad \text{or} \qquad pV = RT \tag{3}$$

where R is the so-called **universal gas constant.** Its numerical value depends upon the units in which p and V are expressed. If p is in atmospheres and V in cubic centimeters, R has the value 82.07 cm³ atm/K for 1 g mol. If p is in pounds per square inch, from now on abbreviated as psi, and V is in cubic feet, R for 1 lb mol is 10.71 psi ft³/R. Several values of R are tabulated in Table 3-1; each is in units of work/mole-degree. (We'll see in the next chapter that the product pV indeed has the dimensions of work.) Note that custom is guilty of using the symbol R for two different quantities. Usually the context will tell you whether it is the universal gas constant as in Equation (3) or whether it stands for the Rankine temperature scale.

Equation (3) holds for 1 mol of any gas. For quantitites other than 1 mol, we write:

$$pV = nRT \tag{4}$$

where n is the number of moles of gas.

We have already remarked that equations like (3) and (4) that relate the temperature, pressure, and volume of a material system are known as equations of state. The one we have developed here, $pV = nRT$, holds for all gases if the pressure is low enough. It is sometimes called the **ideal gas law,** or **perfect gas law.** Any gas obeying it is said to behave as an ideal, or perfect, gas, and will give the "correct" temperature in a gas thermometer no matter what the pressure. For this reason, the gas thermometer scale that we have described is sometimes referred to as the *ideal gas thermometer scale* since all

TABLE 3-1 Gas constant (R) values in commonly used units.*

Pressure (absolute)	Volume	R
Atmospheres (atm)	cubic centimeters (cm³)	82.07 cm³ atm/g mol K
Newtons/square meter (pascals)	cubic meters (m³)	8314 joules/kg mol K or 8.314 joules/g mol K
Pounds/square foot (psf)	cubic feet (ft³)	1545 ft lbf/lbm mol R
Pounds/square inch (psi)	cubic inches (in³)	18,540 in lbf/lbm mol R
Pounds/square inch (psi)	cubic feet (ft³)	10.71 psi ft³/lbm mol R

* See Appendix I for definitions of units and symbols.

gases approach the behavior indicated by $pV = nRT$ as the pressure approaches zero. At high enough pressures, all real gases show substantial departures from $pV = nRT$. Typical nonideal behavior is reflected in the pressure dependence of pV/nRT as shown for several gases in Figure 3-5.

To represent graphically the complete relation between p, V, and T requires three dimensions because there are three variables. Such a three-dimensional plot is a surface on which every point has three coordinate values. Each point on the surface of the earth, for example, is characterized by values for the latitude, longitude, and altitude. A raised-relief map illustrates this three dimensionality on a more comprehensible scale. The surface corresponding to the ideal gas equation of state, $pV = RT$, is shown in Figure 3-6. The dashed lines are isotherms of the kind used to represent Boyle's Law in Figure 3-3. They are simply the curves obtained by the intersection of a constant-temperature plane with the surface. The isotherm of Figure 3-5 is therefore analogous to a contour line indicating elevation when a raised-relief map is reduced to a two-dimensional or flat topographic map.

Figure 3-5 Comparison of real and ideal gas behavior. Only for an ideal gas will $pV = nRT$ at all pressures.

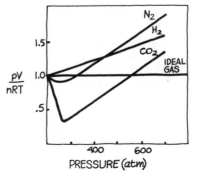

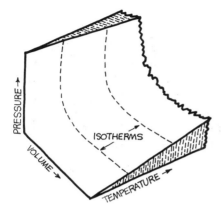

Figure 3-6 A pVT surface for an ideal gas. The dashed lines are isotherms, of which one appears projected on a pV plane in Figure 3-5.

OTHER EQUATIONS OF STATE

Because it is simple and reflects with reasonable accuracy the behavior of many gases over a wide range of conditions, the equation of state $pV = nRT$ is very useful. However, it is hardly universal. Obviously, it will not be obeyed by any matter in liquid or solid form. No such "condensed" material will halve its volume if the pressure is doubled. When they are highly compressed or near liquefaction or condensation points, even gases will show marked departures from $pV = RT$. Many more elaborate equations of state have been proposed. Some of them are highly accurate over a limited range of conditions. Some apply to specific classes of substances. Others are more general with respect to substance and conditions but not very accurate. It is not worthwhile to engage here in any extensive examination of such equations, but it will be useful to develop some perspectives.

Let's assume that the molecules of a gas are perfectly elastic, hard spheres, so small that their aggregate volume is negligible with respect to the volume of the container. We also assume that there are no attractive or repulsive forces between the molecules and that they are in completely chaotic motion, colliding at random with each other and the walls of the container. If we apply elementary classical mechanics to this gas model, we arrive at the relation $pV = RT$ without any reference to empirical generalizations such as the laws of Boyle and Charles. In other words, what we have called an "ideal" gas behaves just as if it were made up of very small, hard spheres that do not interact except when they collide. The pressure exerted by such a gas on any surface is simply the steady-state rate of momentum transfer by the molecules to the surface when they collide with it. When a molecule of mass m hits a surface with a velocity component v_a perpendicular to the surface, and departs with a velocity component v_d, the net momentum transfer, according to mechanics, is the product $m(v_a + v_d)$. These velocities are pretty high (a few hundred meters per second in air), so the impact time is very short and the transfer of momentum very abrupt. But the collisions are so numerous (a few times 10^{23} per square centimeter per second in air at atmospheric pressure) that, as measured by any gauge the pressure appears perfectly steady and continuous. In fact, most direct measurements and observations of gases seem to indicate that they are continuous fluids. That they must comprise large populations of individual molecules is pure inference.

We know from experience that real gases don't conform to the ground rules of the ideal model just sketched. At low enough temperatures and high enough pressures, any gas will condense into a solid or liquid that is relatively incompressible; thus, the aggregate volume of the molecules is not always negligible relative to the volume of the container. Clearly, there are also attractive forces between the molecules that at low enough temperatures can bind them together in condensed form. These observations suggest that one way to arrive at an equation of state more general than the ideal gas law, $pV = RT$, would be to take into account the finite volume of real molecules and the attractive forces between them.

Accounting for the molecular volume is straightforward, at least on a qualitative basis. We simply recognize that the free volume available for molecules to move around in is less than the volume V of the gas by an amount b that relates to the size of the molecules and is sometimes called the *covolume*. Thus, we replace V in the ideal gas law by $(V - b)$ and obtain:

$$p(V - b) = RT \tag{5}$$

This relation is sometimes known as the **Clausius Equation of State** in honor of Rudolf Clausius, the German physicist who played a major role in the development of thermodynamics. We will learn more about his work in a later chapter. Note that as written, Equation (5) is for 1 mol of gas. For n moles: $p(V - nb) = nRT$.

To account for the attractive forces between molecules is a bit more difficult. A molecule in the center of the gas, far away from the container wall, will "see" equal numbers of molecules in all directions. Therefore, the attractive forces are the same in all directions and balance each other out so that there is no net force. When a molecule *approaches* the container wall, it will see more molecules behind it than in front of it. As a result, there is a net attractive force toward the center of the gas. The molecule is somewhat restrained and does not hit the wall as hard as if there had been no attractive force. Because a gas's pressure is due to the transfer of momentum by molecules colliding with the container walls (or any surface exposed to the gas), the pressure exerted by attractive molecules is somewhat less than the pressure exerted by nonattracting molecules. It turns out that this pressure deficit is proportional to the square of the gas density. Therefore, we can write:

$$p = p_i - a\rho^2 \tag{6}$$

where ρ is the density in moles/unit volume, p_i is the pressure exerted by an ideal gas of nonattracting molecules, and a is the proportionality constant characterizing the attractive force for a particular molecular species. We recall that $\rho = n/V$ where n is the number of moles. With some rearranging, Equation (6) can be restated for 1 mole, where a has a characteristic value for a particular gas as:

$$p_i = p + a/V^2 \tag{7}$$

The right-hand side of Equation (7) represents a "corrected" ideal gas pressure that can replace p in $pV = RT$. If we make both the volume correction indicated in Equation (5) and the attraction according to Equation (7), we obtain for 1 mole of gas:

$$(p + a/V^2)(V - b) = RT \tag{8}$$

This expression was first proposed by the Dutch physicist J. D. van der Waals in 1873. For n moles it becomes: $(p + a(n/V)^2)(V - nb) = nRT$.

The van der Waals equation embodies in simple and lucid form the two effects that cause real gases to depart from ideal behavior. A surface representing the van der Waals equation of state clearly cannot be as simple as one for an ideal gas. Part of such a surface for particular values of a and b is represented in Figure 3-7. Isotherms are indicated by solid lines. Those at temperatures higher than that of the so-called *critical isotherm*, labeled T_c, show no dips or inflections and look like the ideal gas isotherms shown in Figure 3-6. At temperatures below T_c, the isotherms show maxima and minima. Indeed, at sufficiently low temperatures, there is a region in which p actually becomes negative as indicated by the dashed portions of the isotherms. These bumps and dips and regions of negative pressure are simply artifacts of the van der Waals equation—deficiencies in its ability to describe the true equilibrium behavior of real substances. What actually happens in

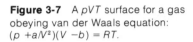
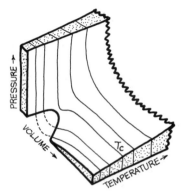

Figure 3-7 A pVT surface for a gas obeying van der Waals equation: $(p + a/V^2)(V - b) = RT$.

real gases is that at temperatures below T_c, where p is sufficiently high, the attractive forces between molecules condense the gas into a liquid or solid. Thus, the anomalous region of peaks and valleys in the isotherms and the negative pressure that the van der Waals equation predicts become, in real substances, a mixed phase region in which vapor and liquid or solid coexist. Figure 3-8 represents this situation. Such "discontinuous" behavior simply cannot be described by any relatively simple and "continuous" equation.

Despite its deficiencies, the van der Waals equation is useful as a correction to the ideal gas law. Values of a and b for various gases have been deduced from experimental observations. Some typical examples are shown

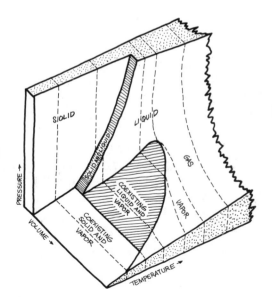

Figure 3-8 A pVT surface for a substance that contracts when it freezes. Surfaces like this one cannot be described by an equation of state and must be constructed from experimental data.

TABLE 3-2 Representative values of van der Waals constants.

Gas	a (cm^6 atm/g mol^2) × 10^{-6}	b (cm^3/g mol)
Helium	0.034	23.7
Methane	2.253	42.8
Water	5.464	30.5
Nitrogen	1.390	39.1
Oxygen	1.360	31.8
Argon	1.345	32.2
Carbon dioxide	3.592	42.7

in Table 3-2. Unfortunately, for any particular gas, no single choice for a and b is sufficient to make the van der Waals equation an accurate description of pVT behavior over a wide range of conditions. Nonetheless, the values in the table do give us some qualitative information on the extent to which departure from ideality can be expected.

It is instructive to examine a particular case and compare results obtained from the ideal gas law, the Clausius equation, the van der Waals equation, and actual measurements. We assume 1 g mol of water vapor in a volume of 1384 cm^3 at a temperature of 500 K. Recalling that R has the value 82.07 cm^3 atm/g mol K and noting from Table 3-2 that $a = 5.464 \times 10^6$ cm^6 atm/g mol^2 and $b = 30.5$ cm^3/g mol, we obtain

(a) from the ideal gas law:

$$p = RT/V = 82.07 \times 500/1384$$
$$= 29.64 \text{ atm}$$

(b) from the Clausius equation of state:

$$p = RT/(V - b) = 82.07 \times 500/(1384 - 30.5)$$
$$= 30.31 \text{ atm}$$

(c) from the van der Waals equation of state:

$$p = \frac{RT}{(V - b)} - \frac{a}{V^2}$$
$$= \frac{82.07 \times 500}{(1384 - 30.5)} - \frac{5.464 \times 10^6}{1384}$$
$$= 27.46 \text{ atm}$$

(d) from experimental measurements:

$$p = 26.07 \text{ atm}$$

For these particular conditions, the ideal gas law is about 14 percent too high. The Clausius equation is even more in error, about 16 percent too high. The van der Waals equation is about 5 percent too high. It is interesting that the Clausius equation is more in error than the ideal gas law. The reason is that the correction for finite molecular volume increases the pressure, whereas the attractive term decreases the calculated pressure. Thus, these two corrections offset one another. By ignoring both of them, the ideal gas law comes nearer the truth than the Clausius equation, which takes into account only the increase in pressure due to a loss of free volume. At much higher densities, the excluded-volume correction becomes much more significant, and the Clausius equation would be more accurate than the ideal gas law.

In general, for real substances we do not know the explicit relation between p, V, T, and n. For most solids and liquids, we don't even have crude approximations. Nevertheless, we are absolutely sure that some such relation exists for every substance and that the substance knows what it is. A block of aluminum will always precisely and reproducibly occupy a particular volume when the temperature and pressure are given particular values. We indicate this general truth by the mathematical statement:

$$f(p, V, T, n) = 0 \tag{10}$$

This statement asserts the existence of some functional relation between p, V, T, and n that can be expressed in an equation. (If all the terms in such an equation are put on the left-hand side, the right-hand side will clearly be zero.) Such an expression is called an **implicit equation of state.** It implies the existence of some relation between the variables. It also implies that we don't know what the relation is but that the substance does! You can get some appreciation of how complex an equation it would take to describe a real substance over a wide range of conditions by looking at Figure 3-7. It shows the surface for a real substance that contracts on freezing as do most substances (except water). We may not be smart enough to figure out ahead of time what volume a substance will occupy for arbitrary choices of p, T, and n, but we have implicit faith that the substance knows what volume to assume. When put to experimental test, that faith has always been fulfilled. Matter is never schizophrenic!

EQUATIONS OF STATE IN VIRIAL FORM

Often it is convenient to expand the right-hand side of Equation (4) for a single mole in terms of a power series:

$$pV = A + B/V + C/V^2 + \cdots$$

This series is known as an equation of state in *virial form*. The problem becomes to obtain expressions for the coefficients A, B, C, etc. and their dependence upon temperature. The table below shows the values of these coefficients for the equations of state we have been considering. You can verify them algebraically.

Virial coefficient	Ideal gas $pV = RT$	Clausius gas $p(V - b) = RT$	Van der Waals gas $(p + a/V^2)(V - b) = RT$
A	RT	RT	RT
B	0	Rtb	RTb − a
C	0	RTb²	RTb²

Thus, the terms after the first term in the series can be regarded as corrections for departure from ideal behavior. Note that in the limit of very low pressures, the volume V of any gas becomes very large so that the terms after the first approach zero even when B, C, and so on are finite. Therefore, the virial equation approaches the ideal gas law just as real gases approach ideal behavior.

SOME SAMPLE PROBLEMS

To illustrate the uses of equations of state and to give you some exposure to units, we will now work out some simple problems. In each case, for simplicity, we will assume the ideal gas equation of state, $pV = nRT$.

1. Consider a balloon 10 m in diameter filled with hydrogen at a pressure of 1 atm and a temperature of 300 K. What is the mass of hydrogen in the balloon?
   (a) Recall that in $pV = nRT$ when p is in atmospheres and V is in cubic centimeters, R has the value 82.07 cm³ atm/K for 1 g mol, that is, 2 g in the case of hydrogen.

(b) For a sphere $V = \frac{4}{3}\pi r^3$, where V is volume and r is radius. For the balloon, $V = \frac{4}{3} \times 3.14 \times 5^3 = 523.3$ m^3 = 5.233×10^8 cm^3.

(c) From the equation of state $n = pV/RT$, where n is the number of moles (gram), $p = 1$ atm, $T = 300$ K, and $V = 5.233 \times 10^8$ cm^3. So $n = (1 \times 5.233 \times 10^8)/(82.07 \times 300) = 21,254$ g mol $= 42,508$ g, or 42.5 kg, of hydrogen.

2. The lifting, or buoyant, force exerted by a balloon is equal to the difference between the weight of the balloon and its contents and the weight of the air it displaces. Assume that the average molecular weight of the gases in air is 29, the balloon skin weighs 10 kg, and that the pressure and temperature of the ambient air are the same as for the hydrogen in the balloon. Calculate the lifting force of the balloon.

(a) Neglect the volume occupied by the balloon skin. Then the volume, temperature, and pressure of the displaced air are the same as for the hydrogen in the balloon. By Avogadro's Law, the number of moles of displaced air must be the same as the number of moles of hydrogen in the balloon, that is, 21,254.

(b) One g mol of air weighs 29 g. Therefore, the weight of the displaced air is $29 \times 21,254 = 616,366$ g, or 616.3 kg.

(c) The balloon and its hydrogen weigh $42.5 + 10 = 52.5$ kg. The displaced air weighs 616.3 kg. Therefore, the balloon can lift $616.3 - 52.5 = 563.8$ kg, or about 1242 lb.

3. An automobile driver checks the air pressure in his tires before starting a long trip on a hot day. The pressure is 31 psi, and the temperature is 72°F. After several hours of driving, he again checks the pressure and finds it to be 37 psi. (We suppose that these indicated pressures are "absolute." The next chapter discusses the difference between absolute and gauge pressures.) What is the temperature of the air in the tire?

(a) We assume $pV = nRT$, no leaks, and no change in the volume.

(b) Because V, n, and R remain unchanged, we can write, with subscript 1 indicating initial state and subscript 2 the final state:

$$\frac{T_2}{T_1} = \frac{p_2 V_2/nR}{p_1 V_1/nR} = \frac{p_2}{p_1}$$

(c) From (b)

$$T_2 = T_1 \times p_2/p_1$$
$$= T_1 \times 37/31 = T_1 \times 1.194$$

(d) Recall that in $pV = nRT$ the T refers to absolute temperature, or gas thermometer temperature. On the Fahrenheit scale, this absolute, or Rankine, temperature is obtained by adding 459.68 to the Fahrenheit temperature. Thus $T_1 = 531.68$ R and $T_2 = 1.194 \times 531.67 = 634.82$ R. We subtract 459.68 to obtain 175°F as the final temperature of the air in the tire, an increase of 103 Fahrenheit degrees.

The change in tire temperature just calculated is not at all unusual. You can see, therefore, why car and tire manufacturers always recommend measuring tire pressure when the tire is cold, and caution against lowering pressure when a tire is hot. The latter can cause serious underinflation at normal temperatures. Note also that this problem is an exercise in applied gas thermometry!

HIGHLIGHTS

The following statements summarize the important points raised in this chapter:

1. A *system* is that part of the real world that we single out for observation and accounting. It is usually some aggregate of material substance, an object or collection of objects. It may be as simple as a bucket of water or as complex as a power-generating station. *Boundary* is a useful concept in identifying a system. It is a sort of closed geometric surface on one side of which is system, on the other *surroundings*.

2. Quantitative description of a system makes use of *properties*. Properties are quantities like length, volume, mass, pressure, and temperature. They are simply numbers resulting from well-defined measurement operations. The *state* of a system is defined, or prescribed, by a particular set of property numbers.

3. The *units* in which properties and interactions (heat and work) are expressed are based upon arbitrary, but agreed-upon, fundamental standards. For mechanical properties, three systems of units are in use: the Metric System, or cgs system, based on the centimeter, gram, and second; the International System (formerly known as the Rational System, or mks system), based on the meter, kilogram, and second; the British Imperial System based on the foot, pound, and second. (See Appendix I.) Temperature, different in kind from purely mechanical quantities, is measured on various scales as discussed in Chapter 2.

4. The set of property values (in particular p, V, and T) describing the state of a system can sometimes be related by an algebraic expression called an *equation of state*. For some substances, the equation of state is known and can be written. For most it is not.

5. A particularly simple and useful equation of state, known as the *ideal gas law*, can be written as: $pV = nRT$, where R is known as the *universal gas constant*. Its numerical value depends upon the choice of units for p, V, and T, but it always has the dimensions of work/mole-degree. Note that in all equations of state, pressures and temperatures *must* be in terms of their absolute values. Do not use gauge pressures. *Always* use Kelvin or Rankine temperatures, of course with the appropriate value of R.

6. In the equation of state, $pV = nRT$, the amount of gas is indicated by n, the number of *moles*. A mole of substance is a number of specific mass, or weight units (grams, pounds, tons, etc.), equal to the *molecular weight* of the substance. The molecular weight of a species is the weight of 1 molecule (minimum stable aggregate of atoms) relative to the weight of 1 atom of the most abundant isotope of carbon, assigned the value 12.0000.

STATESMANSHIP TEST

1. Refer to a table of atomic weights (see Appendix IV) and determine the molecular weights for the following common molecules:

hydrogen sulfide	H_2S	carbon tetrachloride	CCl_4
hydrogen cyanide	HCN	sodium chloride	NaCl
ammonia	NH_3	ethyl alcohol	C_2H_5OH
sulfuric acid	H_2SO_4	uranium hexafluoride	UF_6
propane	C_3H_8	chlorotrifluoromethane (freon)	$CClF_3$

2. The effective molecular weight of air is 29; that is, 1 kg mol of air is 29 kg. (That is to say that air behaves like a gas whose molecules are 29/12 times the mass of carbon atoms.) At standard temperature and pressure (STP) of 0°C and 1.0 atm, how many kilograms of air are in a room measuring $10 \times 10 \times 3$ m? How many pound-moles?

3. Suppose a gas obeys the equation of state $pV = nRT^2$, where T is the ideal gas thermometer temperature, n is the number of moles, V is volume, p is pressure, and R has the value 10^4 J/kg mol K². What is

the gas density in kg/m³ when the temperature is 27°C, pressure is 10^7 N/m², and the molecular weight is 100?

4. Gas A obeys the ideal gas law, $pV = nRT$, and has a molecular weight of 50. One kg mol of this gas is in a container at a pressure of 3×10^6 N/m² when the temperature is 300 K. Gas B obeys the equation of state, $p(V - nb) = nRT$, and has a molecular weight of 100. At 400 K and 2×10^6 N/m², this gas occupies a volume of 1 m³. Which gas has the greater molar density (n/V)? Which gas has the greater mass density $(m/V$, where $m = nM$ and M is molecular weight)? Assume $b = 0.06$ m³/kg mol.

5. The equation of state for the gas *alphon* is $pV = RT$. For *gammon* it is $p(V - b) = RT$. For *rhodon* it is $(p + a/V^2)V = RT$. Assume that all three gases have the same molecular weight, that $b = 0.05$ m³/kg mol, and that $a = 4 \times 10^5$J m³/kg mol². At a given T and p, which gas will have the highest density?   Which the lowest?

6. A gas obeys the equation of state, $p(V - b) = RT$. At $t = 27°C$ and $p = 10^5$ N/m², it has the same density as an ideal gas of molecular weight 100. If $b = 0.10$ m³/kg mol, what is the molecular weight of the gas?

7. The earth's gravitational field attracts the atmosphere and thus causes a pressure at the earth's surface sufficient to support a column of mercury 76 cm high, as in a barometer. Assume that the radius of the earth is 6400 km, the density of mercury is 13.5 g/cm³, the molecular weight of air is 29, that $pV = nRT$, and that $R = 82$ L atm/kg mol. Ignore the variation of gravitational attraction with altitude and determine:
   (a) The total mass of the atmosphere. (Hint: gravity exerts the same force on a unit mass of air as on a unit mass of mercury.)
   (b) The pressure at the earth's surface if each molecule in the atmosphere had its mass increased so that the molecular weight was 87 instead of 29.
   (c) The pressure the atmosphere would exert in the absence of gravitational effects if it were contained in a hollow sphere having the same volume as the earth and were at 300 K. Recall that $V = 4\pi r^3/3$ and $S = 4\pi r^2$, where V is volume of a sphere, r is its radius, and S its surface area.

8. The cabin volume of an airplane is 5000 m³. At cruise altitude the cabin pressure is 500 torr. At takeoff it was 750 torr. Assume air has a molecular weight of 29 and that the cabin temperature remains constant at 300 K. What is the difference in cabin air mass between takeoff and cruise altitude?

4 Back to Work

Earlier we considered some aspects of the interaction we call work. Now we will examine in more detail what work is and how we measure it. It is important to be able to compute the amount of work that may have been performed or that may be required, no matter how it's done or how it's paid for. We will pay particular attention to the computation of work done by an expanding gas. This process is involved in all practical heat engines, and it is the source of most of the motive power on which our technological society is based.

MECHANICAL WORK IN GENERAL

To Charlie the Caveman, and to most of us today, work, in its most primitive sense, is whatever makes us tired. We know from experience that how tired we get, and, therefore, how much work we do, depends upon how much force we exert with our muscles—and how far we have to exert it in pushing, pulling, or lifting. From this primitive experience has come the quantitative, mechanical definition of work as the product of force and distance. As we observed in Chapter 1, this definition in symbolic form is:

$$W = F \times L \tag{1}$$

where W is the amount of work done when a force of magnitude F is exerted

All tired out, his back is achin'
Strained it bringin' home the bacon.

TABLE 4-1 Three ways to measure mechanical work. It all depends on your ruler and your scales.

A force of	Acting through a distance of	Performs work of
1 dyne (= 1 g cm/sec²)	1 centimeter	1 erg (= 10^{-7} joule)
1 newton (= 1 kg m/sec²)	1 meter	1 joule (= 0.73756 ft lbf)
1 pound (= 1 slug ft/sec²)	1 foot	1 foot-pound (= 1.35582 joule)

through a distance L. If you raise a 10-pound weight through a distance of 2 feet, you will have done $10 \times 2 = 20$ foot-pounds of work. There are other units of force and distance and, therefore, work. They are discussed in some detail in Appendix I, but we can summarize the common ones in Table 4-1.

It is important to note that both force and displacement must be involved if work is to be done. In Equation (1), if either F or L are zero, W is also zero. As we noted earlier, no matter how hard you push on a brick wall or how tired you get from doing so, you do no work if the wall does not move. Similarly, a freely falling object whose motion is not resisted by a restraining force, for example, by means of a rope over a pulley, does no work no matter how far it falls, that is, no matter how great the displacement. Note, however, that the gravitational field of the earth performs work on the object; that is, gravitational force accelerates it at a rate of 980 cm/sec². If, by means of a rope and pulley, the object were restrained so that it could move slowly downward but not accelerate, gravity would continue to do the same amount of work on it. Furthermore, the object would perform equal work, which could be measured independently as the tension force on the rope multiplied by the distance of descent. That work could be used, via the rope and pulley, for example, to raise a weight. When the object falls freely, the result of the work by the gravitational field is an increase in what we will learn to call the **kinetic energy** of the object, one-half the product of its mass and the square of its velocity $mv^2/2$. Charlie knew that the faster he threw a stone and the heavier it was, the more tired he became. He just never bothered to use a name and a number to describe what his work had done to the stone when it left his hand.

WORK BY EXPANDING GAS

We are going to be greatly concerned with computing the work done by an expanding gas. A particular case is when a gas pushes on a piston moving in a cylinder. This is the work-producing mechanism used in the first effective

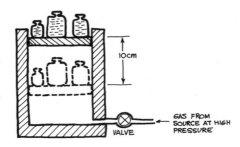

Figure 4-1 An expanding gas can perform work. Here it is raising a piston and weight. It is forced through a throttling valve to maintain a constant pressure on the piston just high enough to push the piston slowly upward.

steam engines and in most automobiles. Let's examine in some detail how we can determine the work done by a gas pushing a piston.

Consider a cylinder containing a piston supported by a gas at some pressure p as shown in Figure 4-1. We will assume that the piston is frictionless and that it is in mechanical equilibrium with the gas, i.e., it is not moving. Thus, the force exerted in an upward direction by the pressure of the gas on the face of the piston equals the force exerted in a downward direction by the weight of the piston and the mass on top of it. Let's suppose that we open very slightly a valve in the line that introduced gas into the cylinder. Gas will then flow very slowly from a source, for example, boiler if the gas is steam. As gas enters the cylinder, the pressure increases. The upward force exceeds the downward force and the piston rises. Suppose we continue to admit gas until the piston has risen a finite distance. Then we shut off the gas, and the piston comes to rest at the position where its downward force again equals the upward force exerted by the gas.

If we know how far the piston has traveled, we can easily compute the work done on the piston. Suppose the upward displacement L is 10 cm, and the mass of the piston and the weight it supports is 8 kg. Recall that in the earth's gravitational field, the force exerted by 1 g against a stationary restraint is 980 dynes. Eight kg would exert a force of $8000 \times 980 = 7,840,000$ dynes. The amount of work is then $980 \times 8000 \times 10$, or 7.84×10^7 ergs $= 7.84$ J (because 1 J $= 10^7$ ergs).

We can also compute the work in another way. Suppose we know the pressure of the gas and the area of the piston. Clearly, the force exerted on the piston will be the product of the pressure and the area. (We assume that we have introduced gas so slowly that the pressure in the cylinder remained constant as the piston moved upward.) In the case we have been discussing, suppose that the diameter of the piston is 10 cm and the pressure of the gas is 10^5 dynes/cm². The force exerted by the gas is the product of the pressure and the area or $10^5 \times 25\pi$ or 7.85×10^6 dynes. The piston moves 10 cm so the work done is 7.85×10^7, or 7.85 J. That is almost exactly the same result that we obtained in the calculation based on the weight of the piston. The value we assumed for the gas pressure was a propitious choice!

In fact, the gas pressure we chose was slightly too high because the upward force exerted by the gas was 7.85×10^6 dynes, a wee bit larger than the downward force of 7.84×10^6 dynes exerted by the piston. If the pressure on the piston were exactly equal in magnitude (but opposite in direction) to the downward force exerted by the piston, there would be no motion and no work. The system with its forces in such a balance would be at **equilibrium,** a state of no motion or change. If the downward force generated by the mass of the piston were only infinitesimally larger than the upward force of the gas pressure, the piston would move downward at an infinitesimally small rate and do work on the gas. Conversely, with only a slight shift in the balance of forces, the direction of motion would be up and the gas would perform work on the piston. Processes occurring when a system is near equilibrium and able to go either way with only infinitesimal changes in driving force, are called **reversible.** The work done during such a process is referred to as **reversible work.** Truly reversible processes are abstractions. They would require infinite time to take place and, therefore, would not really be observable. Charlie would say that nothing was going on. Even so, the concept of reversibility is useful, and we will frequently invoke it as a limiting-case idealization.

In the calculation just carried out, we found that the work was simply the product of the pressure, the area of the piston, and the distance it traveled. But, as Figure 4-2 shows, the product of area and distance is in reality a volume. Thus, we can say that the work done by the gas is equal to its pressure multiplied by its volume change. Because pressure is force per unit area, F/L^2 and volume is L^3, the product pV has the dimensions of work, $F \times L$. Actually, there must be a change in volume if work is to be done. Consequently, to express in symbols what we have said in words:

$$W = p(V_{final} - V_{initial}) = p \, \Delta V \qquad (2)$$

The term ΔV (read "delta vee") is simply a shorthand representation for the volume change—the difference between the initial and final volumes. Very

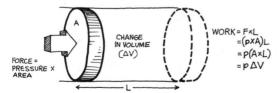

Figure 4-2   An easy way to reckon the work done by gas expanding at constant pressure. Simply multiply the pressure by the volume change.

often it is convenient to consider that a finite amount of work done by an expanding substance is the sum of a large number of infinitesimal amounts of work each associated with a very small volume change. Thus, $\delta W = p\, dV$ where we understand by δW an infinitesimally small, or *differential* amount of work and by dV a very small or, *differential,* change in volume. (For reasons that will emerge we use the symbol δ to indicate small amounts of interaction and d for small changes in a property.) A consequence of this nomenclature is that work associated with volume expansion or contraction of a system is frequently known as $p\, dV$, or "pee-dee-vee," work.

We can calculate $p\, dV$ work by multiplying the pressure and a volume change even when there is not a simple linear displacement. For example, the work done in blowing up a balloon is simply the pressure of the gas in the balloon multiplied by whatever volume change occurs, no matter what the shape. Note that during the inflation of a balloon, the gas does two things. In addition to overcoming atmospheric pressure, which causes an inward force on the balloon's skin, the gas stretches the skin itself. In the case of the moving piston calculation, the product of gas pressure and piston area was equal to the weight of the piston because we stipulated that the piston was frictionless. If friction had occurred between the piston and the cylinder wall, the gas' work would have been greater than in the frictionless case and could not be computed simply by multiplying the weight of the piston by its upward displacement. Moreover, we assumed that no atmospheric pressure acted on the piston. If there had been atmospheric pressure on the outside of the piston, the upward force exerted by the inside gas would have had to be greater. In fact, the pressure of the gas would have had to be 10^5 dynes/cm^2 *above* the outside atmospheric pressure, which exerted downward force on the piston.

Incidentally, 10^5 dynes/cm^2 is just about one-tenth of the pressure that the atmosphere exerts on all surfaces at sea level. A sea level atmosphere is officially 1.013246×10^6 dynes/cm^2, or 1.013246×10^5 N/m^2 (newtons per square meter or pascals). You will recall that such a pressure is just sufficient

In this act of re-creation
Work—not money—brings inflation.

to support a column of mercury 76.0 cm, 29.921 inches, in height. In the perhaps more familiar English units, it amounts to 14.696 psi.

In any case, because almost any pressure gauge used to measure the pressure inside the cylinder would have also been exposed to the atmospheric pressure on the outside of the pressure-sensitive element, it would have registered the pressure of 10^5 dynes/cm^2 that we used in our calculation. The point is that most gauges show directly the *difference* between the pressure to be measured and some reference pressure, usually that of the ambient atmosphere. Such a pressure difference is known as a *gauge* pressure and is indicated by the letter *g*. For example, pressure of 25 psig means 25 pounds per square inch *gauge* as opposed to 25 psia, 25 pounds per square inch *absolute*. At sea level under normal atmospheric conditions, absolute pressure would be 14.696 pounds per square inch higher than gauge pressure. More generally, we can say that the difference between the gauge pressure and the absolute value of the measured pressure equals the absolute value of the local atmospheric, or ambient, pressure, which means:

$$p_{\text{gauge}} = p_{\text{absolute}} - p_{\text{ambient}} \tag{3}$$

The measurement of pressure is discussed in Appendix I.

If in computing the work done by the gas in the cylinder we take into account the possibility of friction and pressure due to an ambient atmosphere, we would have to say that

Work done by gas = Work due to raising piston
+ Work done in overcoming friction
+ Work done in pushing back the atmosphere

Thus, the *total* work done by the gas is *always* equal to the *total* force it exerts multiplied by displacement, or its *absolute* pressure multiplied by its volume change. The *net* work, over and above that done on or by the ambient atmosphere, equals the product of *gauge* pressure and volume change.

In the last chapter, we developed the idea of an equation of state that describes the relationship between pressure, volume, and temperature for a system. In the case of gases, we noted that the expression, $pV = nRT$, was precisely true for all gases in the limit of zero pressure and was a very good approximation over a wide range of conditions. We have just noted that pV must have the dimensions of work. Therefore, because n is a number of moles, we can now see what we merely accepted on faith before—that R must have the dimensions of work per degree per mole. We noted in the last chapter that R was 82.07 cubic centimeter atmospheres per degree per mol, or cm³ atm/mol K. Suppose we have a piston 10 cm² in area pushed by a gas at 1 atmosphere pressure, 1.013×10^6 dynes/cm². If the piston moves 8.207 cm, the work done by the gas will be $10 \times 1.013 \times 10^6 \times 8.207 = 8.314 \times 10^7$ dynes, or 8.314 J. Also, of course, the product of pressure and volume change is 1 atm $\times$ 10 cm² $\times$ 8.207 cm, or 82.07 cm³ atm. Therefore, if R is equal to 82.07 cm³ atm/mol K, it must also equal 8.314 J/mol K, as Table 3-1 claimed.

WORK WHEN THE PRESSURE CHANGES

In the cases we have been considering, we have been assuming that pressure remains constant during volume change. Thus, it is straightforward to write

$$W = p(V_f - V_i) = p \ \Delta V \tag{4}$$

where we understand that V_i is the initial volume, V_f the final volume, and ΔV the difference between the two. But suppose pressure is changing while volume is changing. How do we carry out the computation in that situation? To answer this question, let's look again at the constant pressure process as depicted in the top illustration of Figure 4-3. The pressure remains constant during the expansion from V_i to V_f, and the work is simply equal to the crosshatched area, $p(V_f - V_i)$. Now let's slice up this area as shown in the bottom illustration. The area of the first slice is $p_1 \ dV_1$, the second is $p_2 \ dV_2$, and so on, where the dV's are the differential widths of the slices. (Remember that each $p_j \ dV_j$ represents a differential amount of work, δw_j.) Thus, the total work is the total area obtained by adding up the areas of the individual slices:

$$W = p_1 \ dV_1 + p_2 \ dV_2 + p_3 \ dV_3 + \cdots + p_n \ dV_n \tag{5}$$
$$= \sum_{j=1}^{n} p_j \ dV_j$$

This last expression takes advantage of some mathematical shorthand. $\sum_{j=1}^{n} p_j \ dV_j$ means simply the summation of all the terms from $j = 1$ to $j = n$.

Now let's get back to the case when p is changing. In Figure 4-4 the curve—an isotherm—represents the relation between p and V for a constant temperature process. For an isothermal, or constant-temperature, expansion

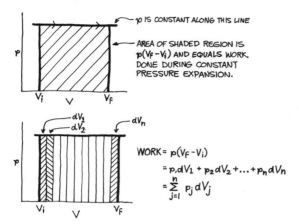

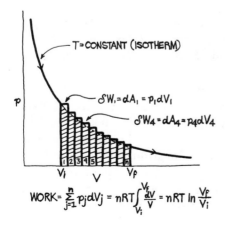

Figure 4-3 Work done by a gas expanding at constant pressure. On a *pV* plot, sometimes called an *indicator diagram,* work is represented by an area, in this case the shaded region under the horizontal line, showing the process (constant pressure) path of the gas.

along the curve shown, work done equals the shaded area under the curve between the initial and final values of the volume. The problem is to compute it. We start by subdividing the total volume change into a number of small volume changes, just as we did in the constant pressure case. Further, we assume that during each of these small volume changes, the pressure remains constant. The result looks like a shelf of books of gradually decreasing height. If we add up the areas, $p \, dV$'s, of the little rectangles, we would have a fair approximation of the total area between V_i and V_f. If we made the dV's smaller, and the rectangles more numerous, we would get a still better approximation. As the dV's become vanishingly small, the sum of the areas of the $p \, dV$'s would equal the area under the curve between V_i and V_f.

But in order to carry out the summation of the $p \, dV$'s as indicated in

Figure 4-4 Work down by a gas expanding at constant temperature. The amount of work is still represented by the shaded area under the curve (isotherm) showing the process path, but it is no longer the product of a pressure and a volume change because the pressure is also changing. To evaluate the work we must resort to integration; that is, we must add up the areas of a large number of very thin slices.

Figure 4-4, we have to know the value of p for each dV. We go to the equation of state for help. For simplicity in computation, we will assume that the ideal gas law holds. If $pV = nRT$ at all times, which is what this law requires, then we can solve for p and find that it equals nRT/V.

$$W = \frac{nRT_1}{V_1} dV_1 + \frac{nRT_2}{V_2} dV_2 + \frac{nRT_3}{V_3} dV_3 + \cdots + \frac{nRT_n}{V_n} dV_n \qquad (6)$$

This equation will hold for any expansion. In the isothermal case we are considering, we can simplify it by noting that all the T's are the same. So we obtain:

$$W = nRT(dV_1/V_1 + dV_2/V_2 + \cdots + dV_n/V_n)$$

$$= nRT \sum_{j=1}^{n} dV_j/V_j = nRT \int_{V_j}^{V_f} dV/V \qquad (7)$$

where we again use some mathematical shorthand.

The symbol $\int$ is called an *integral* sign. It represents the limiting case of Σ when the dV's have become vanishingly small and their total number has become almost infinitely large. The symbols V_i and V_f at the extremities of the integral sign are known as the *limits* of integration. They simply mean that the addition (integration) process is over the interval from the initial volume V_i to the final V_f. Note that $\int$ is reminiscent of the letter S, derived from the Greek letter *sigma* Σ. Sigma Σ is used for any sum, no matter how big the dV's are. The symbol $\int$ *always* relates to the sum of vanishingly small dV's. Whenever the *integrand* (what follows the integral sign $\int$) begins with d, the value of the integral becomes simply the difference between the value of what follows d at the end and the beginning of the interval. Thus, $\int_{x1}^{x2} dx = x_2 - x_1$. (For reasons enlarged upon later, when the integrand is written with δ instead of d, we *cannot* write $\int_1^2 \delta w = w_2 - w_1$.)

The summations, or integrations, represented by Equation (7) are so important and are encountered so often that we will now take a small excursion to see how we can evaluate them. If this excursion is to be very meaningful, you will have to know something about logarithms. For those of you who have previously made their acquaintance, the following boxed review may be useful. Those of you who have not encountered this species of numerical shorthand are urged to take time out for the introduction in Appendix II.

LOG LEXICON

Any number can be expressed as a base number raised to a power indicated by a small superscript number called an *exponent*. The base number is known simply as the *base*. The power to which the base is raised is known as the *logarithm*. Thus, if $y = a^x$, the "logarithm of y to the base a is x" or $\log_a y = x$. One widely used base is the number *10*. Thus $100 = 10^2$, or $\log_{10} 100 = 2$. Logarithms based on the number 10 are known as *common logs*. Tables are widely available. When we write *log* without specifying a base, it is safe to assume that the base is 10.

Another important base is the number $e = 2.71828 \dots$ Like π, the ratio of the circumference of a circle to its diameter, e cannot be expressed exactly without an infinite number of digits after the decimal. In spite of this seemingly awkward characteristic, e like π is very useful. Logarithms based on e are known as *natural* or *Naperian* logs and are usually represented by the symbol *In*. Thus, if $y = e^x$, we say $\ln y = x$.

Because the product of a^x and a^y is a^{x+y}, it is easy to show the following logarithmic properties:

$$\log m + \log n = \log(m \times n) \tag{8}$$

$$\log m - \log n = \log(m/n) \tag{9}$$

$$\log m^r = r \log m \tag{10}$$

These propositions are true no matter what base is assumed.

Sometimes we may have access to only one kind of log table. Therefore, we enquire as to the difference between natural and common logarithms. Suppose $y = 10^r$. Then

$$\ln y = \ln 10^r = r \ln 10 = (\log y)(\ln 10) \tag{11}$$

because $r = \log y$. But $\ln 10 = 2.30258, 40929, 94045, \dots$; so we can write to a fair approximation:

$$\ln y = 2.303 \log y \tag{12}$$

Similarly, $\log e = 0.43429, 44819, 03151 \dots$ and to a similar precision:

$$\log y = 0.4343 \ln y \tag{13}$$

LOG LEXICON   *continued*

Thus, we can obtain one kind of logarithm from another. An important feature of natural logarithms is that when *x* is any very small number (i.e., much less than unity),

$$\ln(1 + x) = x \qquad (14)$$

is a very good approximation. The smaller *x* is, the more nearly true this expression. Clearly, as *x* approaches zero, both sides of the equation go to zero and the expression is precisely true. When *x* has a value as large as 0.1, then $\ln(1 + x) = \ln(1.1) = 0.095$, which is within 5 percent of 0.1.

BACK TO ISOTHERMAL EXPANSION

For our present purpose, the important characteristic of natural logarithms is contained in Equation (14) in the boxed review. We can substitute for a very small number the natural logarithm of 1 plus that number and vice versa. We can make this substitution very usefully in Equation (7), which we will now rewrite for ready reference:

$$W = nRT(dV_1/V_1 + dV_2/V_2 + dV_3/V_3 + \cdots + dV_n/V_n) \qquad (7)$$

Recall that each dV is a very small increment in V. Consequently, the terms dV/V are very small numbers, and we can replace them by log terms in accordance with Equation (14). That is,

$$dV/V = \ln(1 + dV/V) = \ln\left(\frac{V + dV}{V}\right)$$
$$= \ln(V + dV) - \ln V \qquad (15)$$
$$= d \ln V$$

This last step may require a little thought. It says that dV/V, which is simply a very small fractional change in V, is equivalent to $d \ln V$, a small or differential change in the natural log of V. This follows, perhaps not obviously, from the second line in Equation (15), where the right-hand side represents the difference in the natural log of $(V + dV)$ and the natural log of V; that is, the change in the natural log of V as V goes to $V + dV$. We denote this small excursion as a differential change in $\ln V$ or $d \ln V$. If we substitute $d \ln V$ for each of the dV/V terms in Equation (7), we get:

$$
\begin{aligned}
W &= nRT(d \ln V_1 + d \ln V_2 + d \ln V_3 + \cdots + d \ln V_n) \\
&= nRT\Sigma d \ln V_i = nRT \int_{V_i}^{V_f} d \ln V \\
&= nRT(\ln V_f - \ln V_i) = nRT \ln(V_f/V_i)
\end{aligned} \tag{16}
$$

Thus, the work done by an ideal gas expanding at constant temperature is easily calculated by taking the natural logarithm of the ratio of the final volume to the initial volume and multiplying it by nRT, where n is the number of moles of gas, R is the gas constant, and T is the absolute temperature.

The key point of this long and somewhat intricate exercise can be summarized as

$$
\begin{aligned}
\int_{V_i}^{V_f} dV/V = \sum_{i}^{n} d \ln V_i &= \int_{V_i}^{V_f} d \ln V \\
&= \ln(V_f/V_i)
\end{aligned} \tag{17}
$$

The same kind of relation must also hold for any other variable like p or T. Thus we can write:

$$
\int_{T_i}^{T_f} dT/T = \ln(T_f/T_i) \qquad \text{and} \qquad \int_{p_i}^{p_f} dp/p = \ln(p_f/p_i) \tag{18}
$$

This integration of fractional changes in a variable will come up repeatedly. Engrave the following relation in your memory. It will be most useful. Note that x stands for any variable.

$$
\int_{x_1}^{x_2} \frac{dx}{x} = \ln \frac{x_2}{x_1} = 2.303 \log \frac{x_2}{x_1} \tag{19}
$$

Let's close this discussion by carrying out a computation for the work of an isothermal expansion. Suppose we have a piston in a cylinder surrounded by a container of boiling water so the temperature stays at 373 K. We will assume there is 1 g mol of gas in the cylinder. To begin with, the force exerted by the gas in an upward direction is just sufficient to support at a particular level the weight of the piston and an open container of water resting on it as shown in Figure 4-5. As the water in the container evaporates,

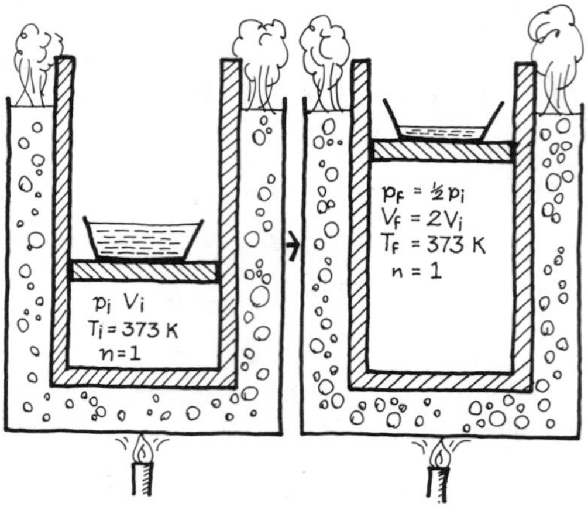

Figure 4-5 One way to carry out a work-performing isothermal expansion.

the downward force exerted by the mass of piston and container will decrease and the piston will rise. Let's assume that it rises until the gas volume reaches twice its initial value. By Equation (15):

$$W = RT \ln(V_f/V_i) = RT \ln 2$$
$$= 8.314 \times 373 \times 0.693$$
$$= 2150 \text{ J}$$

Note that it doesn't make any difference what the actual values of the initial and final volumes are; only their ratio counts. The same amount of work would have been done had the mass of the piston and container been greater, so that the initial volume and final volume for the same amount of gas would have been only half as large. The pressure then would have been twice as great. Doubling the volume at a given temperature means the same amount of work for a particular amount of gas no matter what the pressure. Note that these statements are true only for an ideal gas obeying the equation of state $pV = nRT$. For other equations, the evaluation of $\int_{V_i}^{V_f} p \, dV$ involves the same principles but is a bit more intricate.

It is interesting to compare the amount of work done during this isothermal expansion with the amount that would have been done had we kept the pressure constant. A constant-pressure expansion could have been achieved, for example, by taking the cylinder out of the container of water and putting it on a stove. As the gas warmed up, it would have expanded. If we put a lid on the container of water resting on the piston, the mass would have remained constant and the piston would have risen slowly as the volume of the gas increased at constant pressure. Recalling Equation (4), if the volume doubles, we can write

$$W = p(V_f - V_i) = \frac{RT_i}{V_i}(2V_i - V_i)$$
$$= RT_i = 8.314 \times 373 = 3100 \text{ J}$$

Almost 50 percent more work is done in the constant-pressure expansion than in the constant-temperature expansion. In the case we have been considering, the extra work is done in raising the water that, in the course of the isothermal expansion, evaporated and was not raised by the piston.

EXAMPLE PROBLEMS

At the end of Chapter 3, we computed how much hydrogen was contained in a balloon 10 m in diameter at a pressure of 1 atm and a temperature of 300 K. We found that the volume of the balloon was 523.3 m³, and the amount of gas was 21.25 kg mol, or 42.5 kg. Let's now compute how much work was done by the gas in the balloon as it pushed back the atmosphere during inflation. We will assume the expansion took place at constant ambient pressure of 1 atm, that the final inflated volume is the same, and that we can neglect any work that may have been required to stretch the balloon's skin during inflation.

(a) According to Equation (4), the work done during a constant-pressure expansion is $W = p(V_f - V_i)$;
(b) The initial volume V_i is zero, and the final volume V_f is 523.3 m³. Atmospheric pressure is 1.013×10^5 N/m². Thus:

$$W = 1.013 \times 10^5 \text{ N/m}^2 \times 523.3 \text{ m}^3$$
$$= 530.1 \times 10^5 \text{ Nm (or J)}$$

Recall that we found the balloon could lift 564 kg. Each kilogram exerts a downward force of 9.8 N. Thus, for every meter the balloon lifted a 564 kg payload, it would perform $9.8 \times 564 = 5527$ N m (or J) of work.

If the balloon's lifting ability were independent of altitude, it would have to raise its payload to an altitude of $530.1 \times 10^5/5527$, or 9591 m, in order to perform as much work as was required to blow it up! That is about 700 m higher than Mt. Everest. Actually, the density of the ambient air decreases markedly with altitude, so that the balloon's buoyancy and lifting power would peter out long before it got that high. (However, the balloons used for meterological and astronomical studies have a large amount of slack in their envelopes when they leave the ground; so their volumes can increase as the ambient pressure decreases with increasing height. In this way, their lifting force is maintained to much higher altitudes than would be the case if their volumes were fixed.)

IN BRIEF

The following statements cover the important points brought out in this chapter.

1. *Mechanical work* is performed when a force acts through a distance. The amount of work is determined by multiplying the magnitude of the force by the magnitude of the displacement. Thus:

$$\text{Work} = \text{Force} \times \text{Distance} = F \times L$$

2. The units in which work is measured relate to the units of the force and those of the distance. A force of 1 pound exerted through a distance of 1 foot results in the performance of 1 foot-pound of work. A force of 1 dyne through a distance of 1 centimeter performs 1 dyne-centimeter of work, also called an *erg*. Similarly, a *newton-meter* is sometimes called a *joule*. There are 10^7 ergs in 1 joule.

3. Note that both force and displacement must have nonzero values if work is to have been done.

4. *Pressure* is force per unit area. If we express the units of area as length squared, the units of pressure are F/L^2. The units of volume are L^3. Therefore, pressure $\times$ volume $= F \times L^3/L^2 = F \times L$. Consequently, a pressure multiplied by a volume change is dimensionally equivalent to a force multiplied by a displacement. When a fluid expands, that is, increases its volume, it does work. Frequently, we refer to such work as $p\,dV$ work where dV represents a small change in volume.

5. If pressure remains constant during an expansion, the computation of work done is simply $p(V_{\text{final}} - V_{\text{initial}})$, often written $p(V_f - V_i)$, and sometimes as $p(V_2 - V_1)$.

6. If the pressure changes during the volume change, the computation requires taking account of the pressure change. One way to do this is to assume that the total volume change can be divided into small volume increments, dV. The work done can then be represented:

$$W = p_1\, dV_1 + p_2\, dV_2 + p_3\, dV_3 + \cdots + p_n\, dV_n$$
$$= \sum_j^n p_j\, dV_j = \int_{V_{\text{initial}}}^{V_{\text{final}}} p\, dV$$

where the term $\sum_j^n p_j\, dV_j$ means the sum of all the terms no matter what the size of the dV's. The last term $\int_{V_i}^{V_f} p\, dV$ stands for the sum of all terms over the interval from initial volume to final volume as the dV's approach zero size, that is, the "integral of $p\, dV$ from V_i to V_f."

7. In the case that the fluid obeys the ideal gas law, $pV = nRT$:

$$W = \int_{V_i}^{V_f} \frac{nRT}{V}\, dV = nRT\, \ln(V_f/V_i)$$

if T is constant as during an isothermal expansion. For other equations of state, evaluation of $\int_{V_i}^{V_f} p\, dV$ is somewhat more complicated.

WORK TO LEARN BY

1. Steam at a constant absolute pressure of 20 atm is admitted to the cylinder of an engine. The length of the piston stroke is 60 cm and its diameter is 20 cm. How much work in joules does the steam perform during one full stroke of the piston?

2. Gaseous argon has a molecular weight of 40. How much work will be required to increase the density of 80 kg of argon from 4 kg/m³ to 8 kg/m³ while the temperature remains constant at 600 K? How much would be required if the pressure remains constant at the initial value? Assume $pV = nRT$.

3. One kg mol of steam at atmospheric pressure occupies approximately a volume given by $V = nRT/p$. Water has a molecular weight of 18 and a specific volume in the liquid state of 1 L/kg. If 1 kg mol of water is vaporized at its normal boiling temperature of 100°C, how much work does it do in pushing back the atmosphere?

4. Five kg mol of oxygen occupy a volume of 10 m³ at 300 K. Determine the work required to decrease the volume to 5 m³ (a) at constant pressure; (b) at constant temperature; (c) What is the temperature at the end of process (a) and the pressure at the end of process (b)? (d) Sketch both processes on a pV diagram.

5. One kg mol of an ideal gas quadruples its volume at 400 K. If all of the work done during the expansion were used to compress 1 kg mol of another gas that obeys the equation of state $p(V - b) = RT$, what would the final volume of the second gas be if its initial pressure were 10 atm and its initial temperature were 700 K? Assume the pressure remains constant.

6. Compute the work done by 1 kg mol of an ideal gas at an initial temperature of 500 K if its pressure halves at constant temperature. How much work would be done if the pressure were halved along a path for which $p = 3V$? Sketch both processes on a pV diagram.

7. A diver working at a depth of 100 m in water having a density of 1000 kg/m³ exhales 1 L of gas having a molecular weight of 30. Assume that the gas in the resulting bubble stays at a constant temperature of 300 K as it rises to the surface. How much work does the gas perform during the ascent?

8. A cylinder of helium at 300 K contains 0.25 kg mol, initially at 100 atm pressure. How many spherical balloons 30 cm in diameter can be inflated to 1 atm pressure by the contents of the cylinder? How much work will be done altogether in pushing back the atmosphere? Assume the temperature remains contant at 300 K and that the initial pressures are gauge pressures. Neglect any work required to stretch the rubber envelope.

9. An automobile tire has an inside (wheel) diameter of 16 inches and an outside (peripheral) diameter of 28 inches. Assume that a cross section of the tire is circular and that the volume of the tire is equal to the product of this cross sectional area and the circumference of a circle whose diameter is 22 inches. How much work will be required to inflate the tire to a pressure of 29.4 psig on a day when the atmospheric

pressure is 14.7 psia and the temperature is 300 K? How much work will be required to inflate the same tire to 29.4 psig in Denver when the ambient pressure is 12.3 psia and the temperature is 280 K? Assume that inflation occurs slowly enough so that the air temperature remains constant in both cases.

10. An operatic tenor sustaining a high note for 20 seconds exhales 1 L of air (as measured at 0°C and 1 atm, or 760 torr) in an auditorium where the temperature is 293 K on an evening when the atmospheric pressure is 750 torr. As he sings, the pressure inside his lungs is 780 torr. How many watts is he producing in order to push back the atmosphere to make room for his exhalation? Ignore any volume change in his chest cavity and recall that 1 W is equal to 1 J/sec.

For sailing craft in any fleet
The power source is solar heat.

5 More in re Heat

Up to now in our story we have noted that beginning with Charlie our ancestors had for a long time experienced and exploited a number of heat phenomena. By mid-eighteenth century guns had long been preferred weapons, and the steam engine was becoming a significant replacement for muscle power. But explaining what heat really is, and developing a model by which heat interactions could be understood, remained an unsolved problem. In this chapter we will address ourselves to this problem by examining some early ideas about the nature of heat.

HEAT IS A HAPPENING

Some people argue that there is a strong parallel between the development of the human fetus and the evolution of the human species. It can also be argued that the learning processes of the species are similar to those of the individual. Certainly in reconstructing a plausible account of the experience of early *Homo sapiens* with what we now call heat, we drew heavily on our own individual, primitive sense perceptions. As a point of departure for further considerations, we summarize that early experience, enhanced as it has been by the development of thermometry and the concept of temperature.

1. In describing the results of experience with heat using the sense of touch, we say that some objects are hotter than others. The hotter an object, the higher is its temperature.
2. When two objects at different temperatures are brought together, the hot one cools down and the cold one warms up. In other words, their temperature difference decreases and finally disappears as they approach thermal equilibrium.

3. When placed between two objects at different temperatures, some materials inhibit the rate of temperature change. We call these materials thermal insulators.

Let's reflect a bit on what happens when we bring together two objects at different temperatures. What we observe is that the temperature of one goes up and the temperature of the other goes down. Because the same thing happens every time, we can say that there is a *correlation*, or correspondence, between the temperature changes in each object. Whenever there is a one-to-one correlation between an observable change in one system (object) and an observable change in the surroundings (or another system), we assume that there is a relation between the two changes. Indeed, this assumption is an article of faith that is basic to all science.

Sometimes, it is quite obvious that there is an interaction and, what kind it is is also clear. In the figure it is obvious that every time Charlie pulls down inside the system boundary, the weight outside the system goes up. Therefore, there is an interaction—work. But even when we don't know how to explain or describe the connection, we are still convinced that it exists as long as the correlation occurs. If you walk into a room and flip the light switch, you don't need to see the wires behind the wall to be sure of a cause–effect connection and to "know" that flipping the switch caused the light to go on. Indeed, a born-again Charlie encountering the flip-the-switch-on-comes-the-light phenomenon for the first time would be entirely persuaded after a few tries that there was a cause–effect relation even though he had no comprehension of electricity. Cause and effect have often been perceived before they were understood. People realized that there was a relation between the appearance of clouds and the falling of raindrops long before they comprehended the mechanism of precipitation.

Whenever there is a consistent correlation between a change in a system and a change in another system or somewhere in the surroundings, we call what happens between the system and its surroundings or the other system

Archimedes found his fulcrum
Earth will move now if he works some.

an *interaction*. Thus, when two objects at different temperatures are brought together, we know they interact because both their temperatures change. We agree to call this interaction heat. It is easy enough to say that heat is something that happens between two objects at different temperatures. To explain that "something" is more difficult. We are absolutely convinced that it exists even without being able to say anything more than that there is, in an operational sense, a correlation between the temperature changes in two communicating objects. That correlation is the necessary and sufficient evidence of an interaction between the objects.

HOW MUCH IS THE HEAT THAT HAPPENS?

Before examining further the nature of this interaction we call heat, let's note that thermometry provides the means for determining the *amount* of heat interaction taking place, a concept we have not yet addressed directly. At first there was much confusion about the difference between the *hotness* of an object, which we now characterize as its temperature, and the *amount* of heat interaction required to change an object's hotness. Joseph Black laid much of the ground work for distinguishing between the two. In the 1770s at the University of Glasgow, he examined closely what happened when he mixed various quantities of liquids initially at different temperatures. He also examined carefully the results of heat interactions between objects whose temperature changed and between two-phase systems, for example, ice and water, whose temperature remained constant.

He was able to show that temperature is not necessarily, or even usually, conserved during heat interactions. That is, if the temperature rises in one of two interacting systems of equal mass, it does not necessarily go down by the same amount in the second. Indeed, it might not change at all. But Black demonstrated that there is a quantity, *the amount of heat,* that is conserved. These observations invite the idea that heat is something that flows from one system to another, a concept that had wide appeal for many years before it was discarded. It is more appropriate to interpret the term *heat* simply as the interaction that happens between two adjacent systems that are at different temperatures.

Black was the first to distinguish clearly between the temperature of an object and its *heat capacity*. This latter quantity is a measure of the amount of heat interaction required to effect a change of 1 degree on some temperature scale. It is the basis for the quantitative measure of heat. Black also recognized the difference between the *sensible heat* associated with temperature changes during heat interactions (the basis for measurements of heat capac-

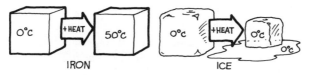

Figure 5-1 Some heat is *sensible* and some isn't. Sensible heat raises the temperature, in this case, that of a block of iron. Heat that affects something without changing its temperature, in this example, melting a block of ice, is called *latent*.

ity) and the *latent heat* associated with phase changes that occur during heat interactions causing no temperature change. (See Figure 5-1.) Therefore, we now call the amount of heat required to melt a unit mass of, say, ice, the **latent heat of fusion** for ice. Similarly, the amount of heat needed to vaporize a unit mass of any substance is the **latent heat of vaporization** for that substance.

In the development of meaningful measurements of mechanical properties such as mass, length, and time, it became necessary to agree on well-defined reference values for these quantities. Establishing a scale to measure heat interactions also required a choice of a reference amount of heat interaction. There emerged two widely used scales, both originally based on water as the reference substance. The **British Thermal Unit,** or BTU, is the amount of heat interaction that raises the temperature of 1 lb of water from 59.5°F to 60.5°F. The **calorie,** abbreviated cal, is the amount of heat interaction that raises the temperature of 1 g of water from 14.5°C to 15.5°C. The temperature intervals must be precisely specified because the amount of heat interaction required to change the temperature of a given mass of water by 1 degree depends slightly on the temperature.

Indeed, there are other definitions of the calorie that differ from the so-called 15° calorie just defined only in the specification of the temperature range of the heating interval. For example, the **mean calorie** is the amount of heat required to raise the temperature of 1 g of water from 0°C to 100°C divided by 100. Similar variations have been associated with the definition of a BTU. At present the calorie and BTU are officially defined in terms of work units. The why and how of such a definition will be discussed later.

Because the calorie is so small, the so-called large calorie is frequently used. It is 1000 times the small calorie and is designated the *kilocalorie*. The usual measure of dietary intake, it is abbreviated kcal or Cal with a capital C. A pound is 454 grams, and a Fahrenheit degree is five-ninths of a Celsius degree. A little arithmetic will satisfy you that a BTU is about 252 small calories.

With these definitions, we can now say more about heat capacity. The **heat capacity** of any object or quantity of a substance is the amount of heat interaction required to raise its temperature 1 degree. The heat capacity for a unit mass is called the specific heat capacity, or **specific heat.** Thus, for water at 60°F, the specific heat is 1 BTU/lb °F; for water at 15°F, it is 1 cal/g°C. Since both of these values are unity (and since they don't change much with temperature), the specific heat of any other material is simply the ratio of the amount of heat required to change its temperature to the amount of heat required to bring about the same temperature change in the same mass of water. Instead of a unit mass of material, it is frequently convenient to refer to a mole (i.e., that number of mass units of a substance equal to its molecular weight). Thus, **molar heat capacity** is expressed in BTU/lb mol °F or cal/g mol °C. On absolute temperature scales, °F and °C can be replaced by R and K.

The specific heat of iron, for example, is 0.11 at room temperature. This means that 0.11 BTU would be required to raise the temperature of 1 lb of iron by 1°F, or 0.11 calories to raise the temperature of 1 g by 1°C. The specific heat of mercury is only 0.033—meaning that the amount of heat interaction that would raise the temperature of 1 g of water by some number of degrees would raise the temperature of 30 g of mercury by the same amount! If you were to conclude on the basis of these two examples that liquid water has an unusually high specific heat capacity relative to other liquid or solid materials, you would be entirely right. Even the specific heat of ice is only half that of liquid water.

It is appropriate to say something about the heat capacity of gases although we are not yet equipped to examine the difference between heating at constant pressure and at constant volume. For diatomic gases such as nitrogen, oxygen, and hydrogen, the heat capacity of 1 g mol at room temperature is nearly the same for each. It is approximately 5/2 R, or 4.97 cal/mol °C in a constant-volume heating process. For hydrogen, this value means that the specific heat capacity is nearly $2\frac{1}{2}$ times that for water. For oxygen, on the other hand, the specific heat capacity is 4.97/32, or 0.155 cal/g °C, only about one-sixth that of water. The heat capacities for constant-pressure heating are somewhat higher than these values for reasons that we will analyze later. They are also more commonly available because they are easier to measure—especially for solids and liquids. Table 5-1 gives some representative values. In general, the molar heat capacities of gases increase with increasing molecular complexity, that is, with the number of atoms per molecule. Because the molecular weight also increases with increasing complexity, the changes in heat capacity per gram are less pronounced.

TABLE 5-1 Representative constant-pressure heat capacities at room temperature.

Substance	c_p (cal/g °C or BTU/lb °F)	C_p (cal/g mol °C or BTU/lb mol °F)
Aluminum	0.215	5.80
Ammonia (gas)	.525	8.94
Argon	.124	4.95
Benzene (liq)	.415	32.4
Benzene (vap)	.249	19.5
Carbon dioxide	.199	8.76
Carbon tetrachloride (liq)	.206	31.7
Carbon tetrachloride (vap)	.129	19.9
Copper	.092	5.85
Ethyl alcohol	.586	27.0
Gasoline (iso-octane)	.494	56.5
Glass (flint)	.117	—
Gold	.031	6.11
Granite	.192	—
Hydrogen	3.41	6.87
Iron	.106	5.92
Mercury	.033	6.62
Nitrogen	.249	6.98
Oxygen	.219	7.01
Salt (sodium chloride)	.207	12.10
Sugar (sucrose)	.299	102.3
Uranium	.028	6.67
Water (ice at 0°C)	.508	9.15
Water (liq)	.998	17.98
Water (steam at 100°C, 1 atm)	.482	8.68

We can summarize these considerations on heat capacity in the simple equation

$$Q = C\,\Delta T = C(T_f - T_i)$$
$$= mc(T_f - T_i) \tag{1}$$

where Q is the symbol for the amount of heat interaction (in cal or BTU) and C represents the total heat capacity of the system whose temperature is changed by ΔT. In simple systems of pure substances, C can be represented by mc where m is the total mass and c the specific heat of the substance. To reiterate: The value of c is unity for water, 0.11 for iron, 0.033 for mercury. For gaseous oxygen at constant volume, it is 0.155. The usual units in each case are cal/ °C or BTU/lb °F. Similarly, C can be taken as nC_m where n is the number of moles and C_m the molar heat capacity with units of cal/g mol °C or BTU/lb mol °F. For liquid water, C_m is 18 cal/g mol °C. Note that lower case c usually refers to unit mass of substance and upper case C to a mole, the subscript m often being omitted.

MODELS AND WHAT CAUSES THAT?

For many purposes and many people, it may be an entirely sufficient description of heat interactions simply to define them operationally as we have thus far done. In other words, what happens is simply what is observed; for example, when two objects at different temperatures are brought together, the temperature of the hotter one decreases and the temperature of the colder one increases until there is no further temperature difference. Such a description is equivalent to explaining the interaction between the earth and an apple by noting that when released from the tree, the apple falls to the ground. Of course, Isaac Newton wasn't satisfied with such a description. Legend has it that a falling apple hit him on the head and triggered the idea of gravitational attraction. Whether the impact was literal or figurative or both, Newton did formulate the law of attraction between masses. (It is interesting to note that Newton also contributed to the development of thermal science. His **Law of Cooling** says that the rate at which the temperature of a body changes is directly proportional to the difference in temperature between the body and its surroundings. It was also Newton who first proposed the temperatures of freezing water and of the human body for thermometric reference points.)

Just as for Sir Isaac it was not enough to assert that apples fall and to say how fast they fall, so it was not enough for some scientists to assert that objects in communication reach the same temperature, and to say how fast and how much the temperature would change. Many such scientists addressed themselves to the question of just what it was that happened when a heat interaction occurred. Clearly, they reasoned, something crosses the boundary of the system whose temperature is changing. Because our individual and collective experience is so absolutely persuasive that there is some physical connection between correlated events, we are entirely convinced that something, real even though unseen, must transcend the boundary of a system undergoing a heat interaction. What is that something? Charlie didn't care about the answer to that question even though heat interactions played an important role in his life. He didn't have time to worry about explanations. He was too busy looking for food and firewood to reflect

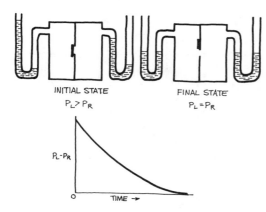

INITIAL STATE
$P_L > P_R$

FINAL STATE
$P_L = P_R$

$P_L - P_R$

O   TIME →

Figure 5-2 Fluids flow from a region of higher pressure to one at lower pressure to equalize the pressures. The larger the pressure difference, the greater the flow rate.

on the mechanism of keeping warm. Only relatively affluent societies can afford scientists and philosophers.

As an explanation of what happens when a body at high temperature has a heat interaction with a body at lower temperature, it was very tempting to draw an analogy with the behavior of a fluid flowing from a container at a high pressure to one at a lower pressure. The higher pressure decreases and the lower pressure increases. The flow continues until the pressure is the same in both containers.

This process is represented schematically in Figure 5-2. The gas pressure is indicated by the manometer fluid. Intuition indicates what experiment would prove—that the rate of flow between the containers depends upon the pressure differences and the cross-sectional area of the connecting aperture or conduit. Further, the rate of flow can be decreased by inserting a fibrous plug, or porous matrix, into the conduit; that is, flow resistance can be introduced. All of these characteristics have their counterpart in a heat interaction between two objects. Temperature assumes the role of pressure. The rate of temperature change in each body depends upon the temperature difference between them and the area of contact or of the cross section of the "conduit" connecting them, for example, a bar of copper. Interaction continues only as long as there is a temperature difference. And the rate of interaction can be decreased by inserting insulators between the two bodies to create resistance. The overall heat interaction process is represented in Figure 5-3. The initial and final temperatures are shown by the height of the liquid columns in the thermometers.

This analogy of heat interaction with fluid flow became very popular and was widely accepted. By the end of the eighteenth century, it had been elaborated into an intricate and comprehensive set of concepts and rules: the

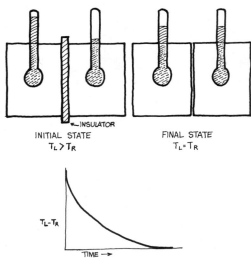

Figure 5-3 A heat interaction between two objects occurs at a rate proportional to the difference in their temperatures and equalizes the temperatures. The analogy to fluid flow is so striking that it was inviting for early investigators to model a heat interaction as the flow of a fluid called "caloric."

Caloric Theory of Heat. Heat was regarded as a fluid called *caloric* that was made up of particles strongly attracted to matter but mutually repulsive. Thus, each atom of matter was surrounded with an atmosphere of caloric. The atoms were attracted to each other with a gravity-like force. Their caloric atmospheres provided a repulsive force. The attractive and repulsive forces just balanced. If more caloric were introduced by heating, the repulsive component of force between the atoms was enhanced and the equilibrium distance increased. Thus, matter expanded when its temperature was raised, that is, as more caloric was added. If matter was compressed so as to decrease interatomic distances, caloric would be squeezed from between the atoms like water from a sponge and the temperature would go up. An object's temperature indicated the escaping tendency or pressure of its caloric—only indirectly the amount of caloric it contained.

Latent heat effects like the melting of ice were ascribed to a chemical reaction between matter and caloric. The temperature would not change until the reaction was over. Careful weighing of objects at different temperatures and of ice before and after melting shows no detectable difference in mass. Therefore, caloric was weightless. Because the same amount of heat was released in cooling a body as was absorbed in heating it through the same temperature interval, caloric was a conserved fluid. It could not be created or destroyed.

The model's plausibility and its qualitative explanation of many thermal effects made it most attractive. Even more persuasive was its ability to provide *quantitative* descriptions of many thermal processes. All of Fourier's work in formulating the theory of heat conduction was based on the concept of heat transfer as a flow of the caloric fluid. In fact, much of the mathematical structure of modern thermodynamics was taken from caloric theory. It is not surprising, therefore, that the caloric theory was accepted by many important scientists including Lavoisier, Laplace, Priestley, DuLong and Petit, Fourier and Clapeyron. But for all its adherents, there were many dissidents. For example, Descartes, Boyle, Bacon, Hooke, and Newton all believed, but could not prove, that heat was somehow identified with the motion of particles making up matter. In fact, this *dynamical* theory of heat, which followed intuitively from the generation of heat by work through friction, was older than the caloric theory but its track record wasn't as good. It had been unable to provide the simple interpretations and quantitative predictions that had been squeezed out of the caloric theory.

THE CALORIC CANON AND RUMFORD'S BORING MILL

One of the most vigorous opponents of the caloric theory was a fascinating character named Benjamin Thompson, alias Count Rumford (see the box). He once said, "I am satisfied that I shall live a sufficiently long time to have the satisfaction of seeing caloric interred with phlogiston in the same tomb."

COUNT RUMFORD

Benjamin Thompson was born in Woburn, Massachusetts, in 1753, of a moderately well-to-do farming family. He was a precocious student of science and medicine with a good grasp of mathematics. At the age of 13 he was apprenticed to a storekeeper in Salem where he carried out chemical and mechanical experiments and learned engraving. A few years later at the beginning of American–British hostilities, he moved to Boston. At the age of 19 he married the wealthy widow of a British colonel, Benjamin Rolfe. She was 14 years his senior. Able then to move in important circles, he wangled a commission as major of a local militia regiment. Because he was suspect to American patriots, prudence arranged his dispatch to England as a messenger. Once there, he obtained an appointment in the Colonial Office and within four years had become an undersecretary of state. Meanwhile, he had maintained his interest and activity in science and in 1779 was elected a Fellow of the Royal Society. After a change in administration, he left the civil service and returned to America. He retired in 1783 with the rank of lieutenant colonel.

Having decided to join the Austrian army in a campaign against the Turks, he met Prince Maximilian and was invited to join the civil and military service

COUNT RUMFORD   *continued*

of Bavaria. With the blessings of the British government, including a knight-hood from George III, he settled in Munich for 11 years as Minister of War, Minister of Police, and Grand Chamberlain to Maximilian. He was an adminis-trator of no mean ability and completely reorganized the army. Something of a pioneer in social reform, he did much to improve the lot of the working classes and to rehabilitate beggars and other social misfits. In 1791 he was made a Count of the Holy Roman Empire. He chose the title of Rumford from the name of the American township (later Concord) from which his wife had come.

He visited England in 1795 but was soon recalled to Munich because of the threats posed by an Austrian and a French army. Rumford's efforts saved Munich from military occupation. He was proposed as Bavarian ambassador to England but his British citizenship created problems, so he returned to London as a private citizen. In 1798 he was instrumental in organizing the Royal Institution and chose Humphry Davy as lecturer in chemistry. In 1804 he left London for Paris and an ill-fated venture into matrimony with Lavoisier's widow. After the marriage ended, he moved to Auteil, on the outskirts of Paris. He died suddenly in 1814 at the age of 62.

Rumford emerges as a flamboyant and free-wheeling operator. He founded, and was the first recipient of, the Rumford Medal of the Royal Soci-ety. He established the Rumford Medal of the American Academy of Arts and Sciences and was donor of the Rumford Professorship at Harvard. Though he was undoubtedly capable, there was always an aura of opportunistic self-interest if not of scandal about his activities. His departure from London to live in France may not have been entirely voluntary.

Whatever his political ethics, Rumford maintained an active interest in science and technology throughout his career. He was an experimentalist and much concerned with practical applications of scientific knowledge. The ex-plosive force of gunpowder and the construction of firearms and signal sys-tems for use at sea particularly interested him. He invented the ballistic pendu-lum and the drip coffeepot. To solve the problem of smoking fireplaces, he developed principles of fireplace design followed to this day. He installed one of the first central heating systems in modern times at the Royal Institution. Rather obsessed with the conviction that the caloric theory of heat was wrong, he was always concerned with thermal phenomena. He developed tech-niques for making weighings accurate to 1 part in a million to demonstrate that heating an object resulted in no measurable weight change.

Phlogiston, you will recall, was a fluid that was once seen as the essence of fire and chemical reaction. It was released during the oxidation of a metal, for example, and absorbed when the metal was reduced to elementary form. Phlogiston's role in explaining chemical reactions was analogous to caloric's role in explaining heat interactions. Antoine Lavoisier founded modern chemistry by exploding the phlogiston theory with his experiments on the oxidation of mercury. Rumford tried to follow his footsteps in more ways

If Rumford's cannon off should go,
He'd see defeathered one scared crow!

than one. Late in life he wed Lavoisier's widow but the marriage did not last long.

Rumford's best-known communication, "Enquiry Concerning the Source of Heat which is Excited by Friction," was presented to the Royal Society in 1798. It reported the famous cannon-boring experiment. With a dull tool in the boring mill, Rumford succeeded in generating enough frictional heat in $2\frac{1}{2}$ hours to boil $26\frac{1}{2}$ pounds of water into vapor. Only 4145 grains (about 9 ounces) of shavings were produced. Because the specific heat capacity of these shavings was the same as the original bulk metal, Rumford argued that the heat could not have been generated by squeezing, or abrading, caloric from the metal. Actually, this argument is a non sequitur. The total heat *content* of a substance may be entirely independent of its heat *capacity,* that is, the amount of heat required to change its temperature. Advocates of the caloric theory could simply argue that the shavings *contained* less heat than the bulk metal, regardless of their relative specific heat capacities. To be convincing, Rumford would have had to show that the bulk metal could be reconstituted from the shavings without an apparent "absorption of caloric," that is, without extraction of heat from some outside source.

With the perspective of hindsight, many people often credit Rumford's work as the beginning of the end of the caloric theory of heat. At the time, however, its impact was less than earthshaking. The absence of weight change upon heating was dismissed as meaning only that the caloric fluid, like electricity, had so little mass as to be experimentally imponderable. Its weightless character was no handicap to the theory. That almost any amount of heat could be generated by friction merely indicated that there was a very large amount of the fluid originally associated with the material atoms. As in the case of frictional generation of static electricity, the fractional amount of fluid released during the rubbing was negligible. Thus, far from being considered as undermining evidence, Rumford's results were welcomed by adherents of the caloric theory as explanations and elucidations of some of its hazier details! With similar equanimity, the caloric theory school accepted the results of Humphry Davy, one year later in 1799, who showed that rubbing two pieces of ice together could cause melting.

In a sense, the Davy experiment was more unequivocal than the Rumford cannon boring. On the basis of careful calorimetric work by Joseph Black and others, everybody agreed that the heat content of water was greater than that of ice. If caloric was conserved in Davy's experiment, why did the ice melt? The caloric theory offered no satisfactory answer. Even so, instead of expiring when confronted with the Davy–Rumford results, the caloric theory reached its peak of development there*after*. Its elegant mathematical formalism evolved in a series of papers by Laplace and Poisson. One famous result was Laplace's correction of Newton's expression for the speed of sound. Newton had computed the velocity of sound in a gas assuming that the compressions and expansions created by the passage of a sound wave occurred at constant temperature. The value he obtained was about 15 percent lower than the experimental value. Laplace assumed that the compressions and expansions caused the temperature of the gas to change. His equation was remarkably close to the experimental value. The differences between these sound-speed models will have more meaning to you after you have read further. Meanwhile, we note also that Fourier's classical, analytical treatment of the conduction of heat in solids appeared in 1822, and Carnot's analysis of ideal heat engines was published in 1824. Both of these classic papers came long *after* the Davy–Rumford experiments.

These important studies and many others after Rumford and Davy assumed the caloric fluid model for heat interactions. There is evidence that Carnot had some misgivings about the model, especially with respect to whether caloric fluid was conserved. Indeed, it seems highly likely in light of his unpublished notes that he had grasped the real nature of heat and its equivalence to work. But it remained for an English brewer's son, James Prescott Joule, to demonstrate the exact equivalence of heat and work interactions and to put the last nail in the caloric theory's coffin. Before we take up Joule's famous experiments, however, we will consider in the next chapter the contribution of his predecessor from across the channel, Sadi Carnot.

EXAMPLE EXERCISES

1. A 5-lb rock at 400°F is dropped into a container holding 9 lb of water at 40°F. The specific heat capacity of the rock is 0.20 BTU/lb °F. Assume no heat loss by evaporation or to the container walls, and compute the final temperature.

(a) Suppose the final temperature is t_x. Then the heat "loss" by the rock is

$$C_{rock}(400 - t_x) = (5 \text{ lb} \times 0.20 \text{ BTU/lb °F})(400 - t_x)°$$
$$= 1.00(400 - t_x) \text{ BTU}$$

(b) The heat "gain" by the water is similarly

$$C_{water}(t_x - 40) = (9 \text{ lb} \times 1.00 \text{ BTU/lb °F})(t_x - 40)°$$
$$= 9(t_x - 40) \text{ BTU}$$

(c) The heat lost by the rock equals the heat gained by the water:

$$9(t_x - 40) = 1.00(400 - t_x)$$
$$10t_x = 760$$
$$t_x = 76°\text{F}$$

Thus the hot rock made the water go from cool to tepid.

2. The heat of fusion of ice is about 80 cal/g. That is to say, 80 cal of heat interaction are required to melt 1 g of ice. Suppose that three 100-g ice cubes at 0°C are dropped into 500 ml of tea initially at 20°C in a thermos bottle. How much ice would melt? Assume that the tea is weak enough that its specific heat is the same as that of water, i.e., 1.00 cal/g °C.
(a) Suppose all the ice melts. That process would absorb 300 g × 80 cal/g = 24,000 cal.
(b) If 24,000 cal were extracted from 500 g of water, the temperature change would be given by:

$$24,000 \text{ cal} = C_{water}(t_i - t_f)$$
$$= 500 \text{ g} \times 1.00 \text{ cal/g} (t_i - t_f)°$$
$$= 500(20 - t_f) \text{ cal}$$
$$500t_f = 10,000 - 24,000$$
$$t_f = -14,000/500 = -28°\text{C}$$

Clearly, there would be ice left because t_f cannot be below 0°C.

(c) Therefore, let M be the number of grams of ice melted.

$$80 \text{ cal/g} \times M = (1.00 \text{ cal/g °C})(500 \text{ g})(20°C - 0°C)$$
$$M = 10{,}000/80 = 125 \text{ g of ice melted}$$
$$300 - 125 = 175 \text{ g of ice left}$$

SPECIFICS ON HEAT

What follows is what's worth recalling from this chapter.

1. *Heat* is an *interaction* between a *system* and its *surroundings* caused by a temperature difference between them. It usually brings about a decrease in that temperature difference.

2. The amount of heat interaction Q between a system and its surroundings is

$$Q = C(T_{final} - T_{initial})$$

where C is the *heat capacity* of the system, T its temperature, and Q is in units based on the heat interaction required to raise a unit mass of water through a unit change in temperature. A British Thermal Unit, BTU, is the amount of heat necessary to raise the temperature of 1 pound of water by 1 degree on the Fahrenheit or Rankine temperature scales. A calorie, or cal, is the amount of heat necessary to raise 1 gram of water through a temperature change of 1 degree on the Celsius or Kelvin temperature scales. A BTU is equivalent to about 252 cal.

3. The *specific heat capacity* of any substance is the amount of heat necessary to raise the temperature of 1 gram of that substance by 1 degree Celsius (or Kelvin) or 1 pound of that substance by 1 degree Fahrenheit (or Rankine.)

4. Sometimes a heat interaction does not change the temperature of a system but changes the amount of liquid, solid, or vapor. The amount of heat required to melt a unit mass of substance is called the *latent heat of fusion*. The amount of heat required to vaporize a unit mass of substance is called the *latent heat of vaporization*.

5. The *Caloric Theory of Heat* was an attempt to represent heat interactions as the flow of a weightless, but conserved, fluid from a system at high

temperature to a system at a lower temperature. The fluid was known as *caioric*.

6. Benjamin Thompson, later Count Rumford, was the most famous of the early antagonists of the now-discredited caloric theory. His cannon-boring experiment, reported in 1798, is one of the most renowned experiments in the history of science. Oddly enough, it was not very influential, either when it was carried out or in the perspective of subsequent history.

WARMING-UP EXERCISES

1. One kg of water is contained in an insulated vessel. Half a liter of mercury at a temperature of 400 K is poured in. The density of mercury is 13.5 g/cm^3, and its specific heat is 0.033 cal/g K. If the initial temperature of the water is 20°C, what is the final temperature inside the vessel after thermal equilibrium is reached?

2. A thermos contains 2 quarts of coffee that was just beginning to boil when it was poured in. Twenty 50-g ice cubes at a temperature of 5°F are dumped in. The specific heat of ice is 0.5 cal/g K, and its heat of fusion is 80 cal/g. Describe the contents of the jug after thermal equilibrium is established.

3. Granular materials like sugar, salt, and sand are not very good heat conductors; so it is difficult to determine their temperatures accurately. Therefore, it is difficult to determine their heat capacities. One way around this difficulty is to disperse the granular material in a liquid in which it is insoluble and to stir it to keep it in suspension—thus assuring a uniform temperature that is easily measured. One then measures the amount of heat required to change the temperature of the mixture and subtracts the amount of heat required to bring about the same change in the pure liquid. The difference is the amount required to raise the temperature of the granular material. In such an experiment, 200 g of carbon tetrachloride are put into a well-stirred, well-insulated calorimeter. It is found that 412.5 cal of heat input are required to raise the temperature of the calorimeter and its contents from 31°C to 42°C. Then, 52.3 g of granular material are added. It is found that 496.6 cal are required to bring about the same temperature rise. What is the specific heat capacity of the granular material?

4. An iceberg weighing 10^8 kg and at a temperature of $-10°C$ drifts into the Gulf Stream whose temperature is 20°C. What is the minimum

amount of Gulf Stream water required to melt the iceberg? If twice that much water became involved, what would be the final temperature, assuming no interaction with the atmosphere or any additional water? (See the second problem in this section for needed data.)

5. Ten kg of ice at $-40°C$ are placed in an insulated container. Five kg of steam at 200°C are introduced. Assume the specific heat of steam is 0.4 cal/g K and the heat of condensation is 540 cal/g. Describe the contents of the container after all changes have occurred.

6. One hundred g of compound X at a temperature of 60°C are dropped into 1 kg of water initially at 20°C and in an insulated container. The final temperature in the container is 21°C. Assume that X is insoluble in water and does not react with it. What is the specific heat capacity of the substance in BTU/lb °F?   in cal/g K?

7. A man owns an extremely well-insulated cabin in northern Maine. Late in January when the temperature is $-40°F$, he comes in on his snowshoes and turns on the electric heater. The cabin warms up to 68°F. Assume that the volume of the cabin is 1000 m³, the original density of the air is 1.4 kg/m³, and its effective specific heat capacity is 240 cal/kg K. Assume that the pressure in the cabin remains constant because of air leaks to the outside. How much heat was used to warm the air remaining in the cabin? Was more heat than this amount put out by the heater? If so, where did it go?

8. One g mol of methane (natural gas) requires just under 10 mol of air for complete combustion that releases 212.8 kcal of heat. (It would be more correct to say that 212.8 kcal of heat interaction with the surroundings would be required to cool the combustion products to the initial temperature of the methane and air.) Assume that a mixture of 1 mol of methane and 10 mol of air give rise to 11 mol of combustion products with an effective molar heat capacity of 10 cal/mol K. If the initial temperature of the mixture is 300 K and if the combustion process is adiabatic (no heat loss to surroundings), what is the final temperature of the mixture?

9. A material has a specific heat capacity that depends upon temperature in accordance with the equation:

$$C = a - b/T$$

where C is specific heat capacity, a and b are constants, and T is Kelvin temperature. In terms of C, a, b, and the initial temperature T_i, show the amount of heat required to double the temperature from its initial value. $\left(\text{Hint: Note that } Q = \int_{T_i}^{T_f} C\, dT, \text{ and refer to Chapter 4 for evaluation of } \int_{T_i}^{T_f} b\, \frac{dT}{T}\right)$

10. Aluminum has a density of 2.17 g/cm³, and a specific heat of 0.215 cal/g K. Silver has a density of 10.5 g/cm³, and a specific heat of 0.056 cal/g K. At this writing, aluminum costs 66 cents/lb and silver costs $44/oz. Two dollars worth of aluminum at 150°C and $2000 worth of silver at −10°C are dropped into a liter of water at 50°C in a thermos. What is the final temperature inside the jug?

Its action here was pretty rough. Caloric must be potent stuff!

6   The Origins of Cycle Analysis

A giant step forward in understanding the heat-to-work process was taken by a young French army officer early in the nineteenth century. In an elegant but simple analysis of an idealized heat engine, Sadi Carnot set an upper limit on the efficiency of any real heat engine. His conclusions became the basis for a formulation of the fundamental principle we now call the *Second Law of Thermodynamics*. Because of its importance, we will review in some detail this contribution of Carnot to science and technology.

NO TOY, THIS ENGINE

In spite of the confusion and abstract arguments during the eighteenth century about the nature of heat, the concrete development of engines to make heat do work had gone on apace. The one thing people seem willing to exercise themselves for is the pursuit of means to avoid exertion. In the struggle to accommodate to nature, devices to save labor have had the highest priority. If necessity is the mother of invention, laziness must be its father. Whatever the reasons, by the time of Rumford's experiments with a dull boring tool at the end of the eighteenth century, the steam engine was working all over Europe. It had been busy pumping water out of mines for decades and was getting ready to pull trains and push boats. Charlie

Action offset by reaction,
He needs oars or wind for traction!

wouldn't have been much interested in pumping water, but he would have loved a steamship.

Clearly, the extent to which the steam engine or any of its cousins is effective in saving labor depends upon the efficiency with which it produces work from the heat released by burning fuel. If, in order to stoke the fire under the boiler, cutting wood or digging coal takes more muscle power than the engine saves, the game is lost and laziness would rebel. Consequently, the efficiency of engines was and is a burning problem. Since the advent of Savery's first mine pumper in 1698, the steam engine had been greatly improved by the inventions and intuitive insight of men like Newcomen, Papin, and Watts. But the improvements were largely a result of empirical development, trial and error. There was little quantitative science underlying the art of engine design and construction. The question of what upper limit there might be to the amount of work obtainable from a cord of wood or a ton of coal was unanswered if not unasked. The first man to grapple effectively with this question was a young French army officer. He achieved immortality with the publication of one small monograph entitled *Reflections on the Motive Power of Heat and on Machines Appropriate for Developing this Power.* The publication date was 1824, just one year before George Stephenson opened the first public, passenger-carrying steam railway in England.

THINKING IN CYCLES BUT NOT CIRCLES

Carnot's contribution to the subject we now call thermodynamics was his conception and analysis of the **cyclic process.** In contemplating the steam engine, he realized that the essence of what went on consisted in the passage of a quantity of water into the boiler where it was vaporized into steam. The steam expanded in the cylinder, pushing the piston. It then exhausted into the condenser where it became liquid water again.

In those days, engines discarded the condensate. But Carnot realized that the condensed water could be returned to the boiler. Thus, the water had gone through a complete cycle—a series of processes or steps that finally returned it to its initial state. Carnot further grasped the essential fact that for the steam engine to run continuously, it was just as necessary to *reject* heat to the cooling water in the condenser as to *absorb* heat in the boiler from the firebox. In other words, the production of work requires the passage of heat from a high temperature source, the fire, to a low temperature sink, the cooling water. More properly, in view of our insistence that heat is an interaction not a fluid, *an engine must interact with a cold sink as well as a hot source if work is to be continuously produced.*

SADI CARNOT: GREATER SON OF A FAMOUS FATHER

Nicholas Leonhard Sadi Carnot was born in 1796, son of a famous French general, Lazare Nicholas Marguerite Carnot. Lazare Carnot was a leader in the overthrow of the monarchy. After a colorful career in public life, during which he completely reorganized the French army, he fell into disfavor after the Second Restoration and lived the last few years of his life in exile. Trained as an engineer, Lazare Carnot published treatises on geometry and mechanics as well as on the less civil aspects of engineering, military fortifications. He must also have been enamored of the arts because he named his son after a then-popular Persian poet, Sadi. (A younger brother was christened Hippolyte after the son of the Amazon queen of Greek mythology!)

At the age of 16 Sadi Carnot entered the Ecole Polytechnique in Paris, founded only two years before he was born. This institution numbered among its early faculty and students some of the most famous names in science. Lagrange, Laplace, Fourier, Berthollet, Ampere, and Dulong were instructors. Cauchy, Coriolis, Poisson, Gay-Lussac, Petit, Fresnel, Biot, Clapeyron, and Poiseuille as well as Sadi Carnot were students. After two years of study Sadi left in 1814 to take a commission in the Corps of Engineers. Shortly thereafter the monarchy was restored, Lazare was exiled, and Sadi found himself doing chores on garrison duty in the boondocks. He managed a transfer to the general staff but almost immediately retired on half pay and moved to Paris where he studied widely in physics and economics.

He familiarized himself with industry and the organization of its factories. He became an expert on trade and industries throughout Europe. After his father died in exile in 1823, Sadi's younger brother Hippolyte came to Paris, and the two set up housekeeping together. Sadi returned to the army for a short time but resigned permanently in 1828 and devoted himself exclusively to study. In 1831 he came down with scarlet fever. He recovered but a year later caught cholera and died at the age of 36.

Sadi Carnot published only one work, a treatise on the motive power of heat in 1824. Pretty much ignored at first, it was something less than a best seller, but its substance has since assured him recognition and fame in the annals of science and technology. Every student of engineering, physics, and chemistry has grappled with the cycle bearing his name. There is hardly a person now alive whose life has not been affected by the heat engines that Carnot first subjected to quantitative analysis. (Even so, his father Lazare rates three times as much space in the *Encyclopedia Britannica!*)

In writing up his analysis, Carnot couched his argument in terms of the caloric theory of heat. He considered the "fall of caloric fluid" through a temperature interval analogous to the fall of water from a high level to a low one. In each case, the amount of work done was determined by the magnitude of the difference between the source and the sink—in temperature for caloric fluid, in height for water. In these terms, the amount of heat (i.e., caloric fluid) rejected to the low-temperature sink was the same as the amount that had been withdrawn from the high-temperature source. According to the then still generally accepted theory, caloric was a conserved (i.e., indestructible) fluid.

Actually, as we will learn, the model characterizing heat as a conserved fluid finally had to be discarded. Indeed, in the notes published by his brother after his death, it becomes clear that Sadi Carnot recognized that the caloric theory was wanting and had grasped the essential nature of heat according to modern concepts. He may have deliberately put his arguments within the framework of the caloric theory to have them more readily accepted by the great majority of scientists and engineers who believed that heat was a conserved fluid. The net result is that he arrived at some erroneous conclusions that were not rectified for several decades. Nevertheless, Carnot's most important conclusions still stand.

Having recognized the essential cyclic nature of the process by which a steam engine produced work by withdrawing heat from the firebox and rejecting it to the condenser, Carnot introduced a simplified, idealized cycle in which air was the working substance. In this way he avoided the complications introduced by the changes of phase, that is, liquid water into steam and then steam back into liquid water. We will now go through a step-by-step analysis of this idealized cycle in much the same way that Carnot did.

CARNOT'S CELEBRATED CYCLE

Before we go into a detailed analysis of Carnot's cycle, we should note some features of gas behavior that we have not yet singled out for attention. In Chapter 3 we considered the relation between the temperature, pressure, and volume of a gas and found that for many situations this relation was adequately described by the equation $pV = nRT$.

What we have not yet considered is what happens to a gas when it undergoes heat or work interactions. Obviously, if we heat it (in the absence of any work interaction), the temperature will rise. Nor should it surprise anyone who has inflated a bicycle tire with a hand pump that compressing a gas raises its temperature. It follows intuitively, that if compressing a gas heats it up, expanding should cool it off. Such cooling actually occurs but is not so commonly confirmed by direct encounter. If you haven't experienced this phenomenon, accept its reality on faith. You might be reassured, however, by realizing that when atmospheric air expands to "lift itself" to high altitudes, the resulting cooling causes rain or snow.

Of course, it is possible to compress or expand a gas and keep the temperature constant. We have only to allow for a heat interaction during the expansion or compression. Indeed, we did exactly this in Chapter 4 when we immersed the cylinder of expanding gas in boiling water to carry out an isothermal expansion. Even as the gas cooled while pushing against the piston, it was warmed by heat exchange through the cylinder walls. We will have much more to say about expansion and compression of gases with and without heat interactions. For now, all we need to know is that if a gas is compressed without heat interaction, that is, *adiabatically,* its temperature will go up. If it expands adiabatically while performing work, its temperature will go down.

Now let's begin our discussion of the Carnot Cycle by considering a cylinder fitted with a frictionless piston and filled with 1 mol of gas, for example, air, as shown in Figure 6-1. The piston and the cylinder walls are perfect thermal insulators. They allow no heat interaction between the gas and the cylinder walls. The bottom end of the cylinder is a good thermal conductor but in this idealized experiment so thin as to be immaterial with zero heat capacity. We start with the cylinder resting on a block of a good thermal conductor, copper for example, at a temperature T_1. We suppose that the copper block is in contact with some heat source that will maintain it at T_1 no matter how much heat "flows" into or out of it.

Note that we still speak of the flow of heat as though heat were a fluid— the caloric theory dies hard! Today we should say that the temperature stays the same no matter how much heat interaction there is between the block and the cylinder. A heat source that maintains a constant temperature no matter how much heat is withdrawn or added is usually called a **reservoir.** It is an idealization, but we can imagine systems that could provide heat while maintaining a constant temperature. For example, a container of steam in equilibrium with hot water could give up heat or absorb it without changing temperature. As heat is withdrawn, some of the steam changes to liquid water at the same temperature. If heat is added, some of the liquid vaporizes. As long as both liquid and vapor are present and the pressure remains constant, the temperature will not change at all despite heat interactions. Similarly, a mixture of ice and water constitutes an isothermal reservoir that can release or absorb heat without changing temperature. Ice melts if heat is absorbed. Water freezes if heat is released.

The upper left-hand corner of Figure 6-1 shows our cylinder of gas resting on the reservoir at T_1. We assume that it has been there long enough so that

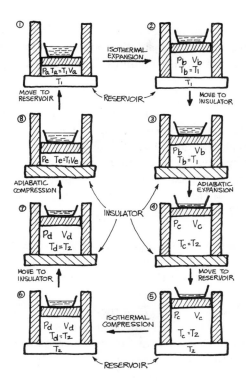

Figure 6-1 An idealized Carnot cycle. Note that the force in the direction of motion is always infinitesimally greater than the resisting force. Thus, the slightest decrease in the driving force or increase in the resisting force would change the direction, so the process is carried out reversibly.

all the gas is also at T_1, the corresponding pressure and volume being designated p_a and V_a. As we start our analysis, the piston has been locked into place. We now release it and allow it to be pushed up slowly by the gas. Actually, the pressure decreases as the expansion occurs because the gas obeys the equation of state, $pV = RT$ (n is unity because we assumed 1 mole of gas in the cylinder), and T is kept constant at T_1 by heat exchange with the reservoir. We can obtain upward motion even with decreasing pressure by allowing water in a container on the piston to evaporate. The resulting weight loss slowly decreases the gas pressure needed to support the piston. We assume that the force exerted by the piston on the gas is always infinitesimally less than the force exerted by the gas on the piston. Thus, the piston will rise very slowly; so we can neglect any velocity effects. Because of the good thermal contact between the gas and the reservoir, the gas temperature is constant at T_1.

We allow this reversible isothermal expansion to continue until the gas reaches p_b and V_b as shown in the part (2) of Figure 6-1. The temperature T_b is the same as T_a, namely, T_1. We now move the piston to the insulator block as shown in part (3) of the figure. Expansion is allowed to continue, but because the gas is now enclosed in a thermally insulating boundary, there is no heat interaction between the gas and its surroundings. The expansion is adiabatic and the gas cools.

The reversible adiabatic expansion goes on until the temperature has dropped to T_2, the temperature of the cold reservoir to which we are about to move the cylinder. At this point, the reaction is represented by part (4) of the figure. The pressure of the gas is p_c, the volume V_c, and the temperature T_c, the same as T_2. We now move the cylinder to reservoir T_2, as shown in part (5) of the figure, where the direction of piston motion changes. We add weight to the piston or push on it so that it starts compressing the gas. Because of the good thermal contact between the gas and the reservoir, the temperature remains constant at T_2.

We continue this reversible isothermal compression until we reach the condition shown in part (6) of the figure. The gas now has pressure p_d, volume V_d, and temperature T_d, the same as T_2. During this isothermal compression, there has been a heat flow from the gas to the reservoir. Now we return the cylinder to the insulating block as shown in part (7) of the sketch and compress the gas reversibly and adiabatically until its temperature rises to T_1, the original temperature. The pressure will be p_e and the volume V_e as shown in part (8). Because the gas obeys the equation of state ($pV = RT$) at all times, the product of pressure p_e and volume V_e must be the same as p_aV_a.

By appropriate choice of termination points in the various steps, we can control the process so that the gas reaches its original pressure and volume as well as its original temperature. That is, $p_e = p_a$ and $V_e = V_a$ where $T_e = T_a = T_1$. Alternatively, when we return the gas to T_1, if p_e and V_e are not the same as p_a and V_a, we then expand or compress the gas isothermally until we reach our target values for p and V. In sum, we can always return the gas precisely to its initial state. We will presume that we have done so in this exercise.

Our description of an idealized Carnot cycle has been fairly long-winded. It is more simply represented with a pV diagram as shown in Figure 6-2. The cycle began with the gas at point a where the pressure was p_a and the volume V_a. The constant temperature expansion then proceeded along the isotherm, $T = T_1$, to point b where the pressure was p_b and the volume V_b. This isotherm is simply the location of all points for which $pV = RT_1$.

At point b, the gas began the adiabatic leg of its expansion. At this stage in our deliberations, we do not know the relation between p and V for an adiabatic expansion, but we do know that the temperature decreases. Therefore, the slope of the adiabatic path must be steeper than for the preceding isothermal process. Otherwise, the temperature would not change. We indi-

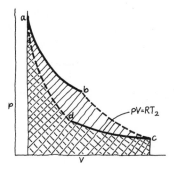

Figure 6-2 The Carnot cycle on a *pV* diagram. The enclosed area corresponds to the net work output during a clockwise traverse.

cate our uncertainty as to its actual path by drawing the adiabatic path as a dotted line in the figure from point b to point c where the temperature has reached the value T_2.

We now begin an isothermal compression along the constant temperature path indicated by the solid line from c to d. At point d, we begin the adiabatic compression along the dotted line from d back to a. If we had chosen the wrong value for p_d and V_d, adiabatic compression might have brought us back to the T_1 isotherm at some point other than a. As we mentioned earlier, this miss would cause no great difficulty. We could proceed along the isotherm by constant temperature compression or expansion until we got back to a. Or we could simply choose whatever point we arrive at as our initial condition. We can always cycle back to that point simply by repeating exactly the steps carried out the first time around. The important objective is to return to exactly the same conditions from which we start. The gas will then have gone through a complete *cycle*.

We now turn to the problem of determining the net results of having carried the gas through this cyclic process. Work was done *by* the gas during the expansion steps. Work was done *on* the gas during the compression steps. Which amount of work was greater? The answer is easy. Inspect Figure 6-2. We learned in Chapter 4 that in a *pV* diagram the area under any curve representing the path of a process; that is, the area between the curve and the *V*-axis, or abscissa, is numerically equal to the amount of work done. Thus, in Figure 6-2, the area under the *ab* and *bc* segments of the path from *a* to *c* represents the work done *by* the system as it expands. This area is identified by solid diagonal lines. As the system is compressed along the path *cda*, the area under the curve is the work done *on* the system. It is identified by dashed diagonal lines. Clearly, the difference between the total work done *by* the system and the total work done by the surroundings *on* the system is the net work done by the system during the complete cycle. This difference, or net work, is the area within the closed curve comprising the segments *ab*, *bc, cd,* and *da.* The ratio of this net work to the heat absorbed during the isothermal expansion from *a* to *b* is called the **efficiency** of the cycle.

Every step around the cycle *abcda* represented in our drawing was *reversible.* An infinitesimal increase in the external force the piston was pushing against during expansion would have changed the expansion into a compression; that is, the surroundings would have done work on the system instead of vice versa. Similarly, during the compression, an infinitesimal decrease in the external force would have changed the compression into an expansion. In other words, we could just as well have caused the system to go counterclockwise around the cycle, *adcba.* In that case, the area under the *adc* part of the closed curve would represent work done *by* the system as it expanded. The area under the *cba* part of the curve would represent the work done *on* the system as it was compressed. And the area within the closed curve would represent the net work done *on* the system.

If you stop to think about it and do a few pencil-and-paper experiments, you will realize that the area within any closed curve on a *pV* diagram represents a net work interaction between a system and its surroundings. If the system goes around the cycle in a clockwise direction, the area represents the net work done *by* the system. If the direction is counterclockwise, the area represents the net work done *on* the system. This conclusion holds for any substance, be it gas, liquid, or solid, no matter what its equation of state. Of course, in the case of solids and liquids, the change of volume with pressure is very small. Consequently, areas enclosed by curves representing any reasonable changes in pressure are very small. In short, such systems don't do much work.

There is another point to be made. As we pursued the system around the cycle in the clockwise direction, it absorbed heat from the high-temperature reservoir at T_1 during the isothermal expansion from *a* to *b.* It rejected heat to the low-temperature reservoir at T_2 during the isothermal compression from *c* to *d.* In the counterclockwise case, the converse will be true. During isothermal expansion from *d* to *c,* heat will be absorbed from the low-

Boiler heat from respiration,
Draggin' tail for cultivation!

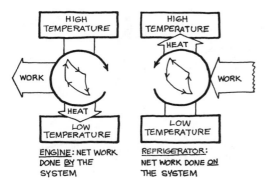

ENGINE: NET WORK
DONE BY THE
SYSTEM

REFRIGERATOR:
NET WORK DONE ON
THE SYSTEM

Figure 6-3 The signs of the heat and work interactions depend upon the sequence of the steps. In a clockwise direction, the cycle produces work as heat flows from the high-temperature reservoir to the low-temperature reservoir. In a counterclockwise direction, the cycle uses work from the surroundings to make heat flow in the opposite direction, from low to high temperature.

temperature reservoir at T_2. Heat will be rejected to the high-temperature reservoir at T_1 during the isothermal compression from b to a.

In sum, the net result of the clockwise cycle is to withdraw heat from the high-temperature reservoir, to reject heat to the low-temperature reservoir, and to perform net work *on* the surroundings. In a counterclockwise cycle, heat is absorbed from the low-temperature reservoir and rejected to the high-temperature reservoir while net work is being done *on* the system *by* the surroundings. When the net result is for heat to flow from the high-temperature reservoir (boiler) to the low-temperature reservoir (condenser) with the performance of work (the clockwise case), we usually say the system is an *engine*. When the cycle is counterclockwise, the heat flow is from the low-temperature reservoir to the high-temperature reservoir, and we call the system a *heat pump* or a *refrigerator* (see Figure 6-3).

CONSEQUENCES OF CARNOT'S CYCLIC CONCATENATION

First a small historical note. In honor of Carnot's contribution, any cyclic process comprising a pair of adiabatic processes and a pair of isothermal processes is now known as a *Carnot cycle*. As we will see, there are a great many different cycles. Some of them bear the names of their inventors or developers: Rankine, Otto, Diesel, Ericsson, Stirling, Brayton. Any of them would have made Charlie dizzy. But the most famous and historically most significant cycle of them all is the one conceived and analyzed by Sadi Carnot.

Carnot's conclusion that the amount of work that could be performed depended directly upon the difference in temperature between the high-temperature source and the low-temperature sink is very important. It is

Figure 6-4 The difference is temperature. The greater the difference between high and low temperatures, the greater the enclosed area and, therefore, the work and the efficiency. To be precise, the efficiency equals $1 - T_{high}/T_{low}$ and is directly proportional to the *ratio* of high and low temperatures.

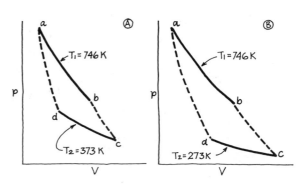

easy to demonstrate in a qualitative way. In part (a) of Figure 6-4 we have the Carnot cycle for 1 mol of gas in which the high-temperature expansion from *a* to *b* occurs at 746 K. The compression from *c* to *d* occurs at 373 K, the normal boiling point of water. In part (b), the high-temperature expansion is also at 746 K. The low-temperature compression is at the freezing point of water, 273 K. Clearly, the enclosed area representing the net work done is larger in the case where the compression temperature is lower. Obviously, the further apart the two isotherms are, the larger will be the area enclosed by the curve representing the complete cycle. (We assume, of course, the same volume changes.) In sum, the greater the temperature difference between source and sink, the greater the amount of work that can be done for each unit of heat absorbed at the high temperature. In other words, the efficiency of the engine is directly proportional to the difference between the temperature at which heat is absorbed and the temperature at which it is rejected.

If the curves in the diagram were accurately drawn, we could measure the area they enclosed. An appropriate scaling factor would then permit us to determine the actual amount of work done. Alternatively, if we have the right equations expressing work in terms of pressure, volume, and temperature, we could calculate the work done by inserting the particular values of the applicable variables. In fact, we learned in Chapter 4 how to compute exactly the amount of work done on or by an ideal gas during isothermal compression or expansion:

$$W = nRT \ln(V_f/V_i)$$

where W is the work, R the gas constant in the ideal gas equation of state, $pV = nRT$, T is the absolute temperature, V_f is the final volume and V_i is the initial volume, n is the number of moles of gas, and ln is the symbol for natural logarithm. (Just to refresh your memory, if $\ln(V_f/V_i) = x$, then $V_f/V_i = e^x$ where $e = 2.71828 \ldots$) With this equation we can calculate the work interactions between system and surroundings during the two isothermal legs of the cycle. We do not yet know how to compute the work done during an adiabatic expansion or compression. Carnot didn't know either; so he too was unable to calculate exactly the total work done while the system went around the cycle. Even so, he arrived at a most important conclusion, one we can also reach by the kind of argument that has become a great favorite with thermodynamicists.

On the basis of what we have said so far, we can imagine the arrangement shown in Figure 6-5. The circle on the right-hand side represents a Carnot engine operating between a high-temperature reservoir T_1 and a low-temperature reservoir T_2. As heat flows through the engine, it produces work used to drive a Carnot heat pump, represented by the circle on the left. Because the only difference between the engine and the heat pump is the direction of the cycle, the corresponding heat and work interactions have the same magnitude and differ only in direction. Thus, the work W, produced by the engine, exactly equals in magnitude the work absorbed by the heat pump. Similarly, the heat interaction Q_1 between the engine and T_1 is exactly the same in magnitude but differs in direction from the heat interaction Q_1' between the heat pump and T_1. Likewise, Q_2 is equal to Q_2'. The net result of a complete cycle by the engine and the heat pump is zero. There will be no net work produced or absorbed. There will be no net gain or loss of heat by either of the reservoirs.

Now let's suppose that we replace the Carnot engine with another engine, X, that operates on a different cycle but still between T_1 and T_2; that is, all heat absorptions are from the reservoir at T_1 and all heat rejections are to the reservoir at T_2. Suppose also that X is more efficient than the Carnot engine. This higher efficiency would mean that for the same amount of heat absorbed for T_1, X would produce more work than the Carnot engine. Thus, it could

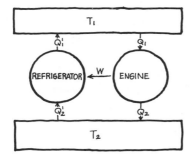

Figure 6-5   A net of nothing. An ideal Carnot engine driving an ideal Carnot refrigerator (heat pump) neither yields nor uses net work and results in no net heat transfer.

drive the Carnot heat pump, which would restore all the heat to T_1, and still there would be some work left over. The net result of such an arrangement would be continuous production of work without any net absorption of heat from the high-temperature reservoir! The excess work could be used to drive another heat pump that would transfer heat from T_2 to T_1. Thus, the overall result would be a spontaneous transfer of heat from a low-temperature reservoir to a high-temperature reservoir. Carnot reasoned that uphill flow of heat was contrary to nature and, therefore, impossible. (Charlie could have told him that!) Consequently, he concluded that no engine cycle could be more efficient than his Carnot cycle.

By analogous argument it is possible to show that no working substance could give rise to a higher efficiency in a Carnot engine or heat pump than any other working substance. Otherwise, the same impossible result would be achieved, namely, perpetual production of work or spontaneous flow of heat from a low temperature to a high temperature. Therefore, all working substances must provide the same efficiency. This corollary to Carnot's conclusion provided the basis for Kelvin's later recognition that a temperature scale could be defined independent of the nature of thermometric substances. If all working substances are equally efficient in a Carnot engine and if the efficiency of such an engine is determined solely by the temperature difference between source and sink, then we could use the efficiency of such an engine as a measure of the temperature difference between two reservoirs. The complete argument is a bit more involved (we shall discuss it in Chapter 10), but it should be intuitively clear that the temperature scale defined in terms of engine efficiencies independent of their working substances will be identical with the ideal gas thermometer scale that we have been using all along.

ENGINES AND THE LAW

It is appropriate to make here a historical observation. Carnot's statement that no combination of engine cycles and working fluids could result in the spontaneous net transfer of heat from low temperature to high temperature is really the first formal assertion of what we now call the **Second Law of Thermodynamics.** The equivalent statements are (1) no engine can have a higher efficiency than one operating in a Carnot cycle; or (2) no engine can generate work without net rejection of heat to a low-temperature reservoir. Note that these statements are not derived from some more general principle. Rather, they represent a generality about the nature of the real world distilled from experience. They are primary truths deserving of the rank of fundamental laws.

It is interesting that what is now known as the Second Law of Thermodynamics was enunciated before that statement of experience we now call the *First Law of Thermodynamics*. In all fairness, however, we should note that the lawlike generality of Carnot's Principle, as it is often called, and all its implications, were not appreciated or completely accepted until after people had come to understand the doctrine of Conservation of Energy, which is the basis for the First Law. The patent offices of the world did not finally refuse to consider inventions that violated the Second Law (i.e., perpetual motion machines of the "second kind" that claimed to convert heat into work without rejection of heat to a low-temperature reservoir) until long after they had refused to consider inventions that violated the First Law (i.e., perpetual motion machines of the "first kind" that claimed to perform work continuously without resort to heat interactions at all). Thus, the First Law was first in finding universal acceptance.

A similar anachronism is involved in the case of the Zeroth Law. As we have noted, this principle asserts the existence of the property called temperature by the statement that if each of two bodies is in thermal equilibrium with a third body, they will be in thermal equilibrium with each other. That this assumption was fundamental to the operational definition of temperature was not recognized until almost 100 years after Carnot made his generalization and 50 years after the First Law had been formulated by Joule, Kelvin, and Clausius. Clearly, however, the concept of temperature and the implied ability to measure it are all assumed in Carnot's work. Consequently, the Zeroth Law is a logical prerequisite to the Second and, as we shall see, First Laws. Hence, its precedence in the usual ordinal enumeration, in spite of its chronological lag.

AN EXAMPLE

Consider an engine operating in an ideal Carnot cycle as shown in Figure 6-2. Assume that the work produced during the adiabatic expansion bc equals in magnitude the work required during the adiabatic compression da. Assume that $V_b/V_a = V_c/V_d = 2$, that there is 1 kg mol of gas in the engine and that the source temperature, T_1, is 746 K. Compute the ratio of net work output, when the sink temperature is 273 K, to the net work output when it is 373 K.

1. The net work output for the cycle will be (assuming $W_{bc} = -W_{da}$):

$$W_{abcd} = W_{ab} + W_{bc} + W_{cd} + W_{da}$$
$$= W_{ab} + W_{cd}$$

2. For an isothermal expansion or compression of 1 mol of ideal gas: $W = RT \ln(V_f/V_i)$. Therefore,

$$W_{ab} = RT_1 \ln(V_b/V_a) = RT_1 \ln 2$$
$$W_{cd} = RT_2 \ln(V_d/V_c) = RT_2 \ln(1/2) = -RT_2 \ln 2$$
$$W_{abcd} = W_{ab} + W_{cd} = (RT_1 - RT_2) \ln 2$$

3. When $T_2 = 273$ K, $W_{abcd} = (746 - 273)R \ln 2 = 473R \ln 2$
   When $T_2 = 373$ K, $W_{abcd} = (746 - 373)R \ln 2 = 373R \ln 2$
   Therefore, $W_{273\text{K}}/W_{373\text{K}} = (473 \ln 2)/(373 \ln 2) = 1.27$

RECYCLE TO REMEMBER

1. A *cyclic process* is any excursion in the property values of a system that returns the system to its original state—the initial set of property values.

2. A *reversible* cyclic process is a cyclic process in which there is no internal mechanical friction, opposing mechanical forces differ only infinitesimally in magnitude, and any heat interactions occur with negligible temperature difference between system and surroundings.

3. Any *heat engine* that operates continuously in a repetitive cycle must reject heat at a low temperature, as well as absorb heat at a higher temperature, if net work is to be performed. If net work is absorbed in a reversible cyclic process, heat will be absorbed at a lower temperature and rejected at a higher temperature.

4. A *Carnot cycle* comprises in sequence an isothermal expansion, an adiabatic expansion, an isothermal compression, and an adiabatic compression.

5. If the direction and sequence of steps is clockwise on a pV diagram of any cycle, heat will be absorbed at higher temperatures, rejected at lower temperatures, and net work will be produced. If the direction is counterclockwise, heat will be rejected at higher temperatures, absorbed at lower temperatures, and net work will be consumed from some external source.

6. The *efficiency* of an engine is defined as the ratio of work produced (what you get) to the heat absorbed from the high-temperature source (what you pay for).

7. No engine can have a higher efficiency than a reversible Carnot engine.

8. All reversible engines have the same efficiency, no matter what the cycle or the working substance.

9. Assertions 7 and 8 are equivalent statements of the general principle known as the *Second Law of Thermodynamics*.

CYCLE DRILL

1. One kg mol of an ideal gas is at an initial pressure of 1 atm and a temperature of 300 K. It triples its volume while the pressure remains constant at 1 atm. Then it expands at constant temperature to 30 times the original volume (i.e., the volume at 300 K and 1 atm).
   (a) Sketch the process on a pV diagram.
   (b) What is the final temperature of the gas?
   (c) What is the final pressure?
   (d) In which step does the gas do more work?

2. One kg mol of ideal gas at an initial temperature T_0, pressure p_0, and volume V_0 goes through the following cycle: (i) It halves its density at constant temperature; (ii) It halves its temperature at constant volume; (iii) It halves its volume at constant pressure; (iv) It returns to its initial condition via a straight line on the pV plane.
   (a) Sketch the cycle on pV, TV, and TP diagrams.
   (b) Determine the work interaction along each leg in terms of R and T_0 (recall $pV = RT$).

3. An ideal gas goes through a Carnot cycle in which the isothermal expansion occurs at 600 K and the isothermal compression occurs at 300 K. At what temperature would the expansion have to take place to double the net work production per cycle? Assume that the volume changes and the sink (compression) temperatures are the same and the net adiabatic work is zero in both cases.

4. A Carnot refrigerator with an ideal gas as a working fluid is extracting heat from a region at 7°C and rejecting it at 27°C. How much more work would be required if the cold region were to be maintained at −13°C?

5. Consider a Carnot cycle in which the sink (low) temperature is T_0 and the source (high) temperature is $4T_0$. Assume that the net adiabatic work is zero. Show that the net work is the same when the gas obeys

$p(V - b) = RT$ as when it obeys $pV = RT$. (Hint: $d(V - b) = dV$ when b is a constant.)

6. One mole of an ideal gas goes through the following cyclic process: (i) An isothermal expansion from V_1 to V_2; (ii) A constant pressure expansion from V_2 to V_3; (iii) An isothermal compression from V_3 to V_4; (iv) A constant pressure compression from V_4 to V_1. Let $V_2 = 2V_1$, and $V_3 = 3V_1$.
   (a) Sketch the process on a pV diagram.
   (b) Determine T_3 and V_4 in terms of initial properties.
   (c) During which legs of the cycle is heat absorbed by the gas?
   (d) In the cycle as described does the system behave as an engine or as a refrigerator? Explain.
   (e) Express the net work interaction during the cycle in terms of R and appropriate initial property values.

7. Assume that a Carnot refrigerator and a Carnot engine each have 1 mol of ideal gas as a working fluid. The engine absorbs heat at 900 K and rejects it at 300 K. During high-temperature expansion, the volume doubles. During low-temperature compression, it halves. The refrigerator absorbs heat at 300 K. Its gas trebles in volume during expansion and decreases to one-third of its initial value during compression. One complete cycle of the engine provides just enough work to drive the refrigerator through one complete cycle.
   (a) Sketch each cycle on a pV diagram.
   (b) At what temperature does the refrigerator reject heat?

Must be prestidigitation.
Else a real hallucination!

7 Heat Is Work and Work Is Heat But Energy's the Difference

A million years or so after Charlie made sticks warm by rubbing them together the exact relation between heat and work interactions began to emerge. History chose a brewer's son to usher in a truly quantitative science of heat. In this chapter, we will learn how his careful experiments on the equivalence of heat and work led to the formulation of the First Law of Thermodynamics.

CHANGE OF SCENE: ARSENAL TO BREWERY

Just four years after the death of Count Rumford and twenty years after the famous cannon-boring experiment, James Prescott Joule was born on Christmas Eve in 1818 at Salford, a small English town near Manchester. Although quite opposite in personality from the flamboyant Rumford, Joule was more successful in hastening the demise of the caloric theory. What he lacked in passion he made up for in precision. Indeed, it was his painstaking accuracy that made his experiments, and the conclusions they led to, so persuasive.

Joule's father owned a large brewery. Its machine shop and other facilities were a great asset to the work that was to become the cornerstone of modern thermodynamics. Because of a spinal injury at birth, Joule was always something of an invalid and was educated at home. Among his tutors was John Dalton, the father of the atomic theory. (After the departure of New College from Manchester in 1799, where he had taught mathematics and philosophy, Dalton had become a "public and private teacher of mathematics and chemistry.") Unlike Dalton, who was a crude experimentalist, Joule was from the beginning convinced of the importance of accuracy. His addiction to quantitative measurement (which Charlie would not have understood!) is revealed in the story that he took a thermometer with him on his honeymoon in the French Alps. He wanted to measure the difference in water tempera-

The work done by this falling weight
Will crack or dent this Newton's pate.

ture between the top and bottom of a waterfall! Why he was interested in such a difference will emerge as we consider his work.

Largely because of Michael Faraday's discoveries, physics in the mid-nineteenth century was dominated by an interest in electromagnetic theory and experiment. Consequently, it is not surprising that Joule's first experiments were concerned with heating effects generated by electric currents from voltaic cells and the chemical reactions taking place in those cells. In the course of this early work, he established the relation that today bears his name. **Joule's law** says that the heat produced by a current I flowing in a resistance R is proportional to I^2R. (More properly, instead of "heat produced," we should say "the amount of heat interaction between a resistor and its surroundings necessary to keep the resistor at a constant temperature.")

Joule then became involved in exploring the relation between mechanical work and heat. A paper in 1843 describes an ingenious experiment in which he ran a generator with falling weights and a system of pulleys. The resulting electric current was passed through a resistor immersed in water. By comparing the temperature rise in the water with the mechanical work done by the falling weights, he determined that 838 ft lb were equivalent to 1 BTU. It is interesting that this experiment also involved an electric current (in keeping with Faraday's influence). In the paper reporting the result, a footnote stated that the more direct method of heating water by forcing it through fine tubes gave a value of 770 ft lb for 1 BTU.

In 1845 Joule reported that measurements of the heat evolved when air was compressed and of the heat absorbed when air was expanded against atmospheric pressure showed 798 as the conversion factor from foot-pounds to BTU's. He also showed experimentally that when compressed gas was allowed to expand freely without performing any work, there was no change in temperature. Then Joule carried out his most famous experiment, one in which falling weights turned a paddle immersed in various liquids, including water, mercury, and sperm-whale oil (see Figure 7-1). The churning caused a temperature increase that Joule measured with a thermometer sensitive to about $\frac{1}{200}$ of a Fahrenheit degree. These measurements gave a value of 782 ft lb/BTU. Some years later he redetermined the mechanical equivalent of heat by again measuring the temperature rise in water surrounding a resistor through which an electric current was passing. The results were so at variance with earlier findings that in 1850 he repeated the weights and paddles experiment with extraordinary care and determined that a BTU was equivalent to 772.5 ft lb. He was thus able to show that the standard of resistance adopted by the British Association (and which he had assumed in his electric heating experiment) was wrong!

Since Joule's experiments, there have been several very careful measurements of the mechanical equivalent of heat by a number of investigators. The most recent deliberate attempt was by Osborne, Stinson, and Ginnings at the National Bureau of Standards in 1939. In terms of the work units that now honor Joule's name, the accepted value is 4.1858 joules per calorie (15°C). If you do the arithmetic, you will find that Joule's final value for the mechanical equivalent of heat, 772.5 ft lb/BTU, corresponds to 4.155 J/cal 15°C—a figure within 1 percent of the present best experimental value! A century of experience in and development of the measuring art has been able to improve on Joule's result by less than 1 part in 100!

More recently, these historically independent units for heat and work interactions have been unified by decree. The New International Steam Table calorie is now defined as exactly 3600/860 J. To five significant figures, this ratio gives the value of 4.1860 J/cal. Similarly, the BTU is now defined as 778.28 ft lb. Defining the calorie in terms of the joule is the final recognition of the equivalence of work and heat. It has practical merit because it removes experimental uncertainties from the relation between the units, both of which are used in reporting a wide variety of experimental data. (A similar

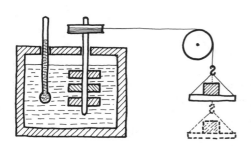

Figure 7-1 To stir will heat (and to heat may stir—by natural convection). A schematic depiction of Joule's epochal experiment.

unification of standards was achieved when the inch was *defined* as exactly 2.54 cm.)

We might ask why Joule's work occupies such an important position in the history of heat. After all, churning up water, whale oil, and mercury isn't all that dramatic and exciting. Indeed, Joule had simply repeated with much less flair the cannon-boring experiments of Rumford. The point is that Joule's work was done so carefully and with such accuracy that the results were accepted almost without question. The full implications of his results took a while to be hammered out, but because the experiments were so persuasive, the conclusions based upon them assumed an aura of infallibility.

Joule's work led ultimately to the First Law of Thermodynamics, a statement that defines a quantity called *energy* and asserts that it is always conserved. The link between Joule's experiments and this important principle will become clearer as we proceed. Joule himself was the first to recognize some of the implications of his experiments, and he formulated the principle of energy conservation as early as 1845. But the first clear formulation of the First Law is usually credited to Rudolf Clausius and William Thomson (later Lord Kelvin) who almost simultaneously in 1850–1851 recognized the importance of Joule's demonstration that heat and work were quantitatively equivalent. Until Clausius and Thomson sorted things out, it had seemed that either Joule was right or Carnot was right, but not both. The dilemma was resolved by the realization that there were two separate principles involved: the First and Second Laws of Thermodynamics.

The First Law follows from Joule's experiments and requires that the heat rejected to the low-temperature reservoir during Carnot's cycle has to be less than the amount absorbed from the high-temperature reservoir. The difference between the heat absorbed and the heat rejected must equal the amount of work produced. The Second Law, which emerged from Carnot's cycle analysis, requires only that there be *some* heat rejected to the low-temperature reservoir, a requirement not implicit in the First Law alone. It does not say how much. Indeed, Carnot's analysis, couched in terms of the caloric theory, assumed that the amount of heat rejected to the sink was the same as the amount absorbed from the source.

You and Charlie might not think this requirement for some heat rejection is terribly important, but it has tremendous practical implications. For example, the oceans can be regarded as virtually infinite sources of heat. Cooling a single cubic mile of sea water by a single Celsius degree would provide all the power needs of the United States for over 24 hours. There are millions of cubic miles of water in the oceans. Moreover, the sun showers heat on them at such a rate that every 10-mile square receives the equivalent of all the U.S. power needs. The trouble is that if we extract heat from the ocean to operate a heat engine, we must dump some of that heat somewhere at a lower temperature. Indeed, according to Carnot's analysis, the heat sink would have to be at a substantially lower temperature if we are to produce an appreciable amount of work. The question is where to find such a sink. One possible answer is at the bottom of the sea. There have been a number of serious proposals to harness the temperature difference between surface water and bottom water, which in some places can be as much as 30 K. A small-scale trial in 1980 off the coast of Hawaii was highly successful.

The other consequences of Carnot's analysis—that efficiency is dependent on the temperature difference between source and sink and that maximum efficiency is independent of working fluid or the details of the particular cycle—are not undermined by the First Law requirement that heat rejected be less than heat absorbed. We will examine the relation between the two laws in more detail in the next chapter when we return to the apparent Joule–Carnot dilemma. Right now we will trace the emergence of the First Law from its empirical antecedents.

Let's reflect on Joule's experiments. Simply, they demonstrated that a particular change in a system could be brought about by either a heat interaction or a work interaction. The temperature of a container of water could be raised either by stirring it or by putting it on a hot stove. Gas could be heated by compression work or by bringing it in contact with a hot object. Of course, this aspect of Joule's work was hardly new. As we have noted, Charlie was well aware that he could heat up a stick either by rubbing it (i.e., with work) or by putting it near the fire (i.e., by heat).

Joule added quantitative accuracy to this knowledge. He was able to assert as a result of his experiments that for any change in a system, the ratio of the amount of work required to the amount of heat necessary to bring about the same change was always identical, that is, 778 ft lb/BTU. More correctly, he showed that the ratio was so nearly the same in a wide variety of systems and for many kinds of work that it was accepted as an absolute and universal truth.

Actually, the equivalence of heat and work has *not* been tested under *all* possible circumstances. But every time careful measurements have been made, the equivalence has been confirmed. More importantly, conclusions derived from assuming equivalence are always found to be true. Thus, we now accept the equivalence of heat and work as an article of faith. We are so sure of this principle that if an experiment seems to show that heat and work are not quantitatively equivalent, then the experiment is suspect.

But what is so great about the equivalence of heat and work? To comprehend its implications, we must look at what was known about mechanics when Joule did his experiments. Most importantly, we must consider what happens when a mechanical system has work interactions with its surroundings.

MECHANICS AND THE ORIGIN OF THE ENERGY CONCEPT

Let's consider a weight of mass m suspended at distance h_0 above the ground. It has the potential for doing work in the course of falling to the ground. For example, we could harness it with a rope and pulley and let it lift another weight. Therefore, we say that, because of its position in the earth's gravitational field, our weight has **potential energy** (PE) equal to mgh_0, where mg is the force exerted by the earth's gravitational field on a mass m. But suppose we don't constrain the weight at all and let it fall freely, doing no work. We would find as Galileo did that it would accelerate at the rate dictated by the constant g, falling faster and faster until it hit the ground. At impact, the weight would no longer have any gravitational potential energy. It would, however, have **kinetic energy** (KE) due to motion in an amount equal to $mv_f^2/2$ where v_f is the velocity at impact. Kinetic energy can also perform work, a notion consistent with the everyday observation that a ball striking a surface will bounce to a height dependent on its velocity and mass (kinetic energy) and its elasticity. Thus the ball's kinetic energy is capable of raising a weight to some height, that is, performing work. (If the ball simply raises itself, of course, there is no work performed because there is no interaction between the ball and its surroundings after it leaves the surface.)

The importance of all this emerges if we idealize the situation. First, we assume that the system of weight and surface is **isolated**, meaning that there are no work or heat interactions between system and surroundings. Second, we take the system to be **conservative**, by which we demand the absence of any frictional or dissipative effects. Now we can say that, at the moment of impact, the kinetic energy equals the original potential energy. Indeed, at any point of the weight's downward plunge, the sum of its kinetic energy ($mv^2/2$) and its potential energy (mgh) is constant:

$$GPE + KE = mgh + mv^2/2 = \text{constant} = mgh_0 = mv_f^2/2 \qquad (1)$$

Now a question may arise. In an ideal, elastic ball bouncing eternally on an ideal elastic surface, where is the energy during the finite time the ball is at rest on the surface? (See Figure 7-2.) The ball is at the bottom of its fall; so it has no gravitational potential energy. It has no velocity; so it has no kinetic energy. However, there is something called *elastic strain energy* (*ESE*), a form of potential energy exemplified by the energy in a compressed spring or a bent bow. For a fleeting moment the total energy of the system is stored in the ball and the surface as the product of stresses (which are forces) and strains (which are displacements) associated with the elastic deformation of the ball and surface upon impact. In the process of springing back to their

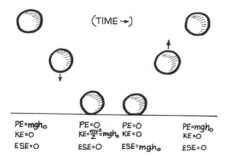

(TIME →)

$PE=mgh_o$
$KE=0$
$ESE=0$

$PE=0$
$KE=\frac{mv^2}{2}=mgh_o$
$ESE=0$

$PE=0$
$KE=0$
$ESE=mgh_o$

$PE=mgh_o$
$KE=0$
$ESE=0$

Figure 7-2 A bouncing ball conserves total energy. *PE* is potential energy (gravitational), *KE* is kinetic energy (due to velocity), and *ESE* is elastic strain energy (like the potential energy in a compressed spring or a stretched rubber band).

original shapes, the ball and surface convert *ESE* to *KE*, which raises the ball back to its original height as *KE* becomes *GPE*.

WORK AND CONSERVATIVE SYSTEMS

For isolated ideal conservative systems in which only gravitational potential energy and translational kinetic energy are involved, we can thus write a slight variation of Equation (1):

$$E = GPE + KE = mgh + mv^2/2 = \text{constant} \tag{2}$$

meaning that if we compare E at two different times (1 and 2)

$$E_2 - E_1 = \Delta E = \text{constant} - \text{constant} = 0 \tag{3}$$

Now, what happens if this system does work on its surroundings? What happens when a weight (the system) falls slowly to the ground, raising—as it falls—another weight by means of a rope and pulley arrangement? At ground level and at rest, the decrease in the total energy of the system would equal the work done in raising the other weight. Thus, the change in total energy of a system equals the work done on or by the system:

$$\Delta E = \Delta KE + \Delta PE = W \tag{4}$$

or

$$\Delta E = \Delta mv^2/2 + \Delta mgh = W \tag{5}$$

Note that we do not use the symbol ΔW for work. Work is not energy. Work is not something *possessed* by a system, whereas energy is. Work is something that *happens* to a system. It is what we have called an interaction. A system *has* energy but it *performs* work. Work is an interaction by which a system changes its energy. We cannot say that a system *contains* so much work, so we cannot represent an amount of work by a change in the "work content" of a system. Thus, ΔW is not used to represent the amount of work done, just W by itself is. Equation (5) says only that the change in energy is numerically equal to the amount of work. It does not say that work and energy are identical quantities.

As we have noted previously, it is often convenient to address ourselves to very small changes in a system. Consequently, we usually use the symbol d to indicate a small or differential change and reserve Δ for large changes. In differential form Equation (5) becomes:

$$dE = d(mgh) + d(mv^2/2) = \delta W \tag{6}$$

where the δ instead of d in front of W reminds us that work is not the same kind of quantity as energy, height, and so on. While d means "a small change in," δ is to be read as "a small amount of." The actual changes in the potential and kinetic energy terms are respectively in h and v^2. So we write:

$$dE = mg\,dh + m\,d(v^2/2) = \delta W \tag{7}$$

Let's examine more closely the quantity $d(v^2/2)$ representing a small change in the square of the velocity, that is, $(v_2^2 - v_1^2)/2$ for cases where v_1 and v_2 are very close together. Recall from elementary algebra that

$$\frac{v_2^2 - v_1^2}{2} = \frac{(v_2 + v_1)(v_2 - v_1)}{2} = \bar{v}(v_2 - v_1) \tag{8}$$

where $\bar{v} = (v_1 + v_2)/2$, the average of v_2 and v_1. Clearly, if v_2 and v_1 are very close, if the change in v is very small, then the difference between $\bar{v}$ and v is negligible; $(v_2 - v_1)$ becomes dv, and we can write:

$$m\,d(v^2/2) = mv\,dv \tag{9}$$

Thus, we can rewrite Equation (7) as

$$dE = mg\,dh + mv\,dv = \delta W \tag{10}$$

Note that velocity v is the ratio of a change in distance to a change in time and can be represented by dx/dt where dx is a small change in distance during the small interval of time dt. Thus,

$$mv\,dv = m(dx/dt)dv = m(dv/dt)dx \tag{11}$$

But dv/dt is a change in velocity with time, or an acceleration. Therefore, $m(dv/dt)$ is a force. Consequently, $mv\,dv$ is equivalent to force multiplied by distance and must have the dimensions of work. Similarly, as we noted earlier, mg is a force and dh is a distance (height change). Therefore, the potential energy term $mg\,dh$ also has the dimensions of work. In sum, energy always has the dimensions of work so Equation (7) is dimensionally consistent. Considerations like these led to the common definition of energy as the ability to do work. It will become clear later that this attractive notion is gravely oversimplified.

MECHANICS EXTENDED

Let's now consider Joule's results in light of this modest excursion into classical mechanics. Suppose we take 1 lb of water as a system. Suppose further that we want to perform 100 ft lb of work on it. We have at least three options:

1. We can raise it 100 ft in the air so that it has 100 ft lb of gravitational potential energy.
2. We can accelerate it to a velocity of about 80 ft/sec where it will have 100 ft lb of kinetic energy. (If 80 ft/sec doesn't seem obvious, recall that the rational unit of mass, which would be accelerated 1 ft/sec² by a 1-lb force, is the *slug*. Therefore, mass in slugs is weight in pounds divided by 32.16 ft/sec², the acceleration due to gravity. Thus, for a 1-lb weight,

$$\frac{mv^2}{2} = \left(\frac{1}{32.16}\right)\frac{v^2}{2} = 100 \text{ ft lb}$$

so $v = (64.32 \times 100)^{1/2} \simeq 80$ ft/sec.)

3. We can churn it with paddles, so that its temperature changes by an amount equaling the temperature change consequent to an equivalent amount of heat. Using the value of 778 ft lb/BTU, we compute the temperature change as 100/778, or about 0.13 degrees on the Fahrenheit scale. (Recall that 1 BTU will raise the temperature of 1 lb of water by 1°F.)

Let's write equations representing these options. For the first two:

$$(1) \quad PE = mg\,\Delta h = W = 100 \text{ ft lb} \tag{12a}$$

$$(2) \quad KE = m\,\Delta v^2/2 = W = 100 \text{ ft lb} \tag{12b}$$

To maintain symmetry, it is inviting to write for (3):

$$(3) \quad IE = C \, \Delta T = W = 100 \text{ ft lb} \qquad (12c)$$

where we have used the abbreviation *IE* for *Internal Energy*, the counterpart of Potential Energy and Kinetic Energy in the first two cases. The observable change in the system is ΔT, C being a proportionality constant that represents the heat capacity of the water.

In terms of this suggestion, we would say that the change in potential energy is reflected in a change in height in the gravitational field. The change in kinetic energy shows up as a change in velocity. The change in internal energy is indicated by a change in temperature. All this makes a very neat package, and so it is very appealing. If true, it would allow us to expand the classical mechanical statement of the conservation of energy to include *internal* energy. Thus, Equation (4) would become:

$$\Delta E = \Delta PE + \Delta KE + \Delta IE = W \qquad (13a)$$

or in terms of differential changes in directly observable properties:

$$dE = mg \, dh \; + \; m \, d(v^2 2) + mc \, dT = \delta W \qquad (13b)$$

where c is the specific heat. Now we can consider that the total energy E includes all three kinds of energy—potential, kinetic, and internal. Therefore, we can still write:

$$dE = \delta W \qquad (14)$$

There is an additional attraction about this idea. Remember that we can write the law of conservation of mechanical energy as

$$PE + KE = \text{constant}$$

only for isolated conservative systems—ones in which there was no dissipation. In such ideal, conservative systems, pendulums swing forever, balls never stop bouncing, and flywheels never slow down. We all know that in the real world frictional dissipation makes all systems displaying observable motion ultimately grind to a halt. But when careful measurements are made on a nonconservative system that runs down, there is always a temperature rise associated with the system's coming to rest.

By taking into account the existence of internal energy, we now have a

If nothing breaks, he's safe, of course,
Sir Isaac's Laws are all in force.

nice explanation of this dissipation. It corresponds to the conversion of potential or kinetic energy into internal energy inside the system; that is, without benefit of a work or heat interaction across its boundary. This conversion is analogous to the conversion of potential energy into kinetic energy as an object falls in a gravitational field. Of course, there is one big difference. The conversion of potential energy into kinetic energy is *reversible* in the sense that kinetic energy can be converted back into potential energy—as when a ball bounces or a pendulum swings. On the other hand, the conversion of internal energy back into kinetic or potential energy has some fundamental limitations that we will look at later on. For now let's content ourselves with the realization that Joule's experiments showed that an internal energy change resulting from a work interaction was exactly equal in amount to the kinetic or potential energy that brought about the work interaction. Consequently, the law of conservation of energy could be extended to include nonconservative systems, thus becoming much more general and far more useful.

You may wonder that it took so long to arrive at this generalization. After all, Charlie rubbed two sticks together and they got hot. The key observation, Joule's contribution, stemmed from the accuracy of his measurements. In many mechanical systems, the temperature rise resulting from running down is quite small. Remember that Joule found it took 778 ft lb of work to raise the temperature of 1 lb of water by 1°F. Thus, 1 lb of water that fell 778 ft, reaching a velocity of about 150 mph, would undergo a temperature rise of only 1°F if all of its kinetic energy were converted into internal energy.

The point is that many mechanical systems such as pendulums and flywheels do not have enough kinetic and/or potential energy to bring about a very large temperature change even if all their mechanical energy ends up as internal energy. And, in many cases, the running down takes a relatively long time and the possible temperature rise, already small, is attenuated by heat exchange with the surroundings. Early observers of mechanical systems didn't bother to insulate them from the surroundings, nor did they carry thermometers in the first place. Perhaps that band of scientists before Joule can be forgiven for not having formulated what now seems so obvious.

AND NOW ADD HEAT

We are not quite through with the implications of Joule's experiments. Equation (13b) points out that by including the temperature of a mechanical system in our bookkeeping we can balance the accounting. We invoke the idea of internal energy to account for the consequences of work interactions that do not show up as changes in kinetic or potential energy. In the limit, if there are no changes in kinetic and potential energy, then all of the work done on a system must be accounted for by an increase in internal energy that shows up as a temperature change or a phase change such as a melting of ice or a boiling of water. But remember that we can change the internal energy of a system by heating it as well as by working on it and that Joule's experiments showed that there was a quantitative equivalence between heat and work. For a given change in a system's temperature, the amount of heat required—as measured in BTU's or calories—equals the amount of work required—as measured in foot-pounds or joules—to bring about the same change in temperature.

Consequently, we can expand our statement of the conservation of energy to include an accounting of heat interactions. Thus, Equation (13a) becomes:

$$\Delta E = \Delta IE + \Delta KE + \Delta PE = W + JQ \qquad (15)$$

where Q is the symbol representing the amount of heat interaction, and J is the proportionality constant Joule established as the ratio of work to heat for a given temperature change in a system. In the case of Joule's "native" units, J is 778 ft lb/BTU. We could express, therefore, the amount of heat interaction Q in terms of foot-pounds if we simply multiplied the value in BTU's by 778. In short, because they are equivalent, we can express both work W and heat Q in the same units and forget about the proportionality constant J. Then the expression for conservation of energy becomes:

$$dE = \delta W + \delta Q \qquad (16)$$

where, as before, d means a small change in the property E and δ indicates a small amount of the interaction Q or W.

Equation (16) constitutes what we now call **The First Law of Ther-**

modynamics. It simply asserts that any differential change in the total energy of a system dE is equal to the sum of the work and heat interactions δW and δQ. Clearly, we can integrate both sides of Equation (16) to obtain the same equivalence for large changes in the system's energy. As written, Equation (16) implies that all quantities have the same sign. That is to say, if δW and δQ are positive, dE will also be positive. A moment's reflection will convince you that in this case, W would have to be work done *on* the system and Q would have to be heat absorbed *by* the system. This choice of signs for Q and W is arbitrary. Usually, W is considered positive when work is done *by* the system just as Q is considered positive when heat is absorbed *by* the system. Using the preposition *by* for both heat and work interactions leads to the more familiar differential statement of the First Law:

$$dE = \delta Q - \delta W \qquad (17)$$

The rationale of these sign conventions may be confusing. Usually, it is intuitively clear just what the signs should be. If a system performs work, its energy must decrease. If it absorbs heat, its energy must increase.

Joule's experiments made possible the extension of the concept of *energy* to include thermal effects. Energy as a conserved quantity had been an extremely useful bookkeeping device in the treatment of mechanical systems. But as long as its definition was limited to purely mechanical potential and kinetic terms, it could only be useful in systems where those terms completely accounted for all observable effects. Thus, in any quantitative sense, mechanics could only treat conservative systems—those in which the total energy, considered as the sum of kinetic and potential energy, was constant. With the concept of *internal energy,* a complete accounting can be done on nonconservative, mechanical systems. Now the total energy of the system includes kinetic, potential, *and* internal energies. Any apparent decrease in the sum of kinetic and potential energies in an isolated system can be accounted for by an increase in internal energy. Note again that by *isolated* we mean that there are no heat or work interactions. Algebraically, we can express isolation by the statement:

$$\delta Q = \delta W = 0 \qquad (18)$$

which is, of course, simply an assertion that there is no change in the total energy of the system or:

$$dE = 0 \qquad (19)$$

(cf. Equation (17).

OTHER KINDS OF ENERGY?

For the systems we have considered we assumed that all kinetic energy is due to translational velocity v and all potential energy is due solely to position in a gravitational field. If there is rotary motion, as in a flywheel, or if there are compressed springs or stretched rubber bands in the system, there would have to be additional terms in any complete description of the system's energy. Similarly, it has been assumed that any changes in internal energy are reflected only in temperature changes. If there are additional possibilities like phase and composition changes, we must include additional terms.

It is inevitable that we ask about effects in addition to those we have called *mechanical* and *thermal*. (And why not? After all, some of Joule's earliest experiments had involved changing the temperature of a system by passing an electric current through a resistor, the current having been produced either by a generator operated by falling weights or by a galvanic cell in which chemical changes were taking place.) For instance, an additional term must be added to the First Law expression when we consider the position of a magnetically susceptible material in a magnetic field. It takes mechanical work to pull a piece of iron away from a magnet, and mechanical work can be performed by harnessing the motion of the iron as it is attracted by the magnet. In this case, the term is entirely analogous to the term for the position of a mass in a gravitational field.

Charges in an electric field represent another case in point. It takes work to bring about displacements resisted by the so-called coulomb forces due to electric charge interaction with an electric field. Electric currents capable of doing work are created by chemical reactions in galvanic cells. Thus, the ability of a system to do work also depends upon its chemical composition. The *total energy* of a system can include contributions from all kinds of kinetic, potential, and compositional effects. Consequently, any change in the total energy E of a system must take into account all these effects; and in the expression for the total differential dE, there must be a term for each component:

$$dE = d(mv^2/2) + mg \, dh + c \, dT + q \, d\epsilon + H \, dM + \mu \, dn \ldots \quad (20)$$

The last three terms represent changes in electric potential, magnetization, and chemical composition.

Often, of course, we can ignore many of these terms because the systems we consider and the processes we treat do not involve all of them.

However, when we simplify analysis by eliminating terms, we must remain alert to the possibility of the intrusion of additional effects. Indeed, we are so absolutely convinced of the generality of the First Law of Thermodynamics, the conservation of energy, that if an apparent exception arises we insist either that the experiment was in error, or we invent a new kind of energy that will balance the books. A classic case of such invention is the **neutrino**. Because they so strongly believed in the conservation of energy, physicists accepted as inevitable the existence of a particle that carried away some of the energy of the nuclear disintegrations that gave rise to beta rays (electrons or positrons). For a long time, there was no direct evidence of the particle's existence. It brought about no directly observable result such as a track in a cloud chamber or on a photographic emulsion. But an energy balance in terms of the known energies of the parent and daughter nuclei and the measured energy of the emitted ray always showed an energy defect. The only feasible explanation, assuming that energy had to be conserved, was that an unseen particle had made off with the missing energy. It was many years before neutrinos, as the great Italian physicist Enrico Fermi christened them, were more directly observed. But their existence was never really in doubt once the energy defect had been established.

EXERCISES

1. The highest waterfall in France is at Gavarnie in the Pyrenees where the overall drop is 1385 feet. If Joule had gone there on his honeymoon instead of to the Alps, what is the maximum difference in water temperature he could have found between the top and the bottom of the falls?

   (a) *What Happens.*   Consider 1 lb of water. Assume no work or heat interactions and no evaporative cooling, and let it come to rest after falling 1385 ft. Because there are no interactions (work or heat) with the surroundings, the total energy of the pound of water remains unchanged. What occurs is the transformation of 1385 ft lb of gravitational potential energy (initially acquired by the work performed in bringing the pound of water to the top of the falls) into internal energy.

   (b) *The Arithmetic.*   The careful measurements of Joule and subsequent workers show that 778 ft lb of work (gravitational potential energy) are equivalent to 1 BTU. One BTU will raise the temperature of 1 lb of water by 1°F. Therefore, the temperature change can be reckoned from:

$$T = \frac{1385 \text{ ft lb}}{778 \text{ ft lb/BTU}} \times 1°\text{F/BTU} = 1.78°\text{F}$$

This temperature rise is large enough to measure on even a fairly crude thermometer. The trouble is that a lot of evaporative cooling can occur during the fall. The heat of vaporization of water is about 1000 BTU/lb. Thus, evaporation of only 0.2 percent of the water would cause a temperature decrease bigger than the increase resulting from conversion of gravitational potential energy into internal energy. No wonder Joule's honeymoon experiment was doomed to fail.

2. A fairly high-powered rifle will shoot a bullet at a velocity of about 500 m/sec. Suppose a lead bullet traveling at this velocity strikes a target and comes to rest without bouncing. Suppose further that the target is an insulator, so the bullet loses no heat. Assume the temperature of the bullet (due to friction with the barrel and contact with the propellant gases) is 100°C before it hits the target. If the specific-heat capacity of lead is 0.022 cal/g °C, its melting temperature is 327.5°C, and its heat of fusion is 26 cal/g, how much, if any, of the lead in the bullet will be melted by the collision with the target?
   (a) The kinetic energy of the bullet is (per gram of mass):

$$mv^2/2 = 1 \times (50{,}000)^2/2 = 25 \times 10^8/2 \text{ ergs}$$

We have expressed v in cm/sec and m in grams, so that the energy is in ergs. There are 10^7 ergs in a joule and 4.186 joules in a calorie. Therefore, the kinetic energy of the bullet is:

$$\frac{12.5 \times 10^8}{10^7 \times 4.186} = 29.86 \text{ cal/g}$$

   (b) To heat the bullet from its initial temperature of 100°C to the melting point of 327.5°C requires:

$$(T_{mp} - T_i)\, c_{Pb} = 227.5 \times 0.022 = 5.005 \text{ cal/g}$$

where c_{Pb} is the specific heat of lead. Thus, of the 29.86 cal/g of kinetic energy only 5.005 are needed to heat the bullet lead to its melting point. Consequently, 29.86 − 5.005 or 24.86 cal/g remain available for melting. The heat of fusion of lead is 26 cal/g. Therefore, 24.86/26 = 0.956, or almost 96 percent, of the bullet will be melted.

THE ESSENTIALS OF EQUIVALENCE

1. Joule's experiments provided quantitative evidence of the equivalence of heat and work. This equivalence means that to bring about a particular observable change in a system, for example, an increase in its temperature by say 10°F, the amount of heat interaction required is *always* in a fixed ratio to the amount of work interaction necessary to bring about the same change. In the metric system, this ratio is 4.1858 J/cal. In the British system, it is 777.9 ft lb/BTU.

2. In mechanics, we find that *energy* is a useful bookkeeping quantity. *Kinetic energy,* associated with the velocity of an object, is expressed as $mv^2/2$, where m is its mass and v is its velocity. Gravitational *potential energy* is given by mgh, where m is mass, g is the acceleration due to gravity, and h is the height in the gravitational field. Another potential energy is the *elastic strain energy* (ESE) associated with the bending of a bow, the stretching of a spring, and the compression of an elastic ball. In addition, there are potential energies associated with various magnetic, electrical, and chemical phenomena.

3. In mechanical systems like bouncing balls and swinging pendulums that are ideal and *conservative,* there is no friction or dissipation and the motion never stops. Often in such systems, energy is repeatedly transformed back and forth from one form to another. For example, with a bouncing ball gravitational potential energy becomes kinetic energy during the fall. During impact, kinetic energy becomes elastic strain energy and then kinetic energy again. During the rebound, kinetic energy becomes gravitational potential energy.

4. In ideal, conservative, mechanical systems, the total mechanical energy remains constant unless there is a work interaction between the system and the surroundings. The system can lose mechanical energy if it does work on the surroundings and gain energy if work is performed on it by the surroundings. If an object is raised in the earth's gravitational field, its potential energy is increased by an amount equal to the work done on

it. That work can be recovered when the object falls to its original position. Considerations like these lead to the appealing but somewhat oversimplified notion that energy is the ability to do work.

5. Joule showed quantitatively that a work interaction can change the temperature of a system without increasing its mechanical (potential or kinetic) energy. This observation led to the idea that a system has *internal* energy associated with its temperature, kinetic energy associated with its velocity, and potential energy associated with its height in a gravitational field.

6. Because of friction, systems in the real world tend to lose kinetic and potential energy. Always associated with this energy loss is an increase in internal energy—usually reflected in a temperature rise. We conclude that the running down of real systems is identifiable with the transformation of mechanical energy into internal energy. For example, a work interaction may raise a ball in the earth's gravitational field. Upon release, the ball bounces up and down as kinetic, gravitational, potential, and elastic strain energy are transformed back and forth. As the height of the bounce gradually decreases, the temperature gradually increases. When the ball finally comes to rest, all of the original gravitational potential energy, equal to the work done in raising the ball, has been transformed into internal energy. The total energy, including internal energy, has remained constant after the initial work interaction. Thus, when internal energy is included in the accounting, all systems can be considered conservative as long as they are isolated. That is, their total energy remains constant in the absence of interactions.

7. The equivalence of heat and work means that heat interactions can also change the internal energy of a system because they also result in temperature increases or decreases (or changes in phase or composition or both). Thus, we arrive at a general equation about energy. For very small changes:

$$dE = \delta q - \delta w$$

Any change dE in the total energy of a system equals the amount of heat δq absorbed by the system less the amount of work δw, performed by the system. This statement is known as the *First Law of Thermodynamics*. Engrave it on your memory as one law you will never break.

EQUIVALENCE EXERCISES

1. A high jumper weighing 160 lb just clears a bar that is 7 ft high. What is his rate of ascent in ft/sec at the instant he leaves the ground if his center of gravity is 3 ft from the ground at takeoff? (Assume that it is his center of gravity that just clears the bar, and recall that in the kinetic energy term $mv^2/2$ if v is in ft/sec, m must be in slugs to give kinetic energy in ft lbf. One slug is close to 32.2 lbm.)

2. A field athlete puts a 16-lb shot in such a way that the maximum height it reaches is 35 ft. If the shot is 6 ft from the ground when it leaves the shot-putter's hand, what is its vertical velocity? Does it have more or less kinetic energy in the vertical direction than the high jumper had in the previous problem?

3. From the top of the Empire State Building (the top of the television antenna) to street level is 1472 ft. The Washington Monument is 555 ft tall. How much faster will a 1-kg weight dropped from the Empire State Building be traveling when it hits the ground, than a 10-kg weight dropped from the top of the Washington Monument?

4. Suppose a meteorite weighing 1 kg and traveling at 6 km/sec buries itself in an iceberg whose temperature is 0°C. Assume that the initial temperature of the meteorite is 4000 K, its specific heat capacity is 0.10 cal/g K, and the heat of fusion of ice is 80 cal/g. How much ice can be melted?

5. When a gallon of gasoline is burned in air, it releases about 140,000 BTU of heat. If 40 percent of that heat can be converted into mechanical work by the engine in a truck, what is the minimum amount of gasoline required to accelerate the truck from a standing start to 60 mph if the gross weight of truck and contents is 40 tons?

6. One kg mol of an ideal gas quadruples its volume at 400 K. If all of the work done during the expansion were dissipated in churning 20 kg of water, how much would the temperature of the water rise?

7. To vaporize 1 g of water requires about 538 cal of heat interaction at 100°C. The resulting vapor at 1 atm occupies a volume of 1674 cm³. What fraction of the heat of vaporization represents the work required to push back the atmosphere to make room for the vapor?

8. In Chapter 3, we noted that attractive forces between molecules "restrained" those at the periphery of a gas, causing them to strike

the boundary with a lower velocity than would be the case in the absence of such forces. In liquids, the same kind of imbalance of forces on surface molecules gives rise to the phenomenon of *surface tension,* which is really a kind of negative pressure. Therefore, to increase the surface area of a liquid, this surface tension must be stretched in the same sense that the envelope of a balloon must be stretched when it is inflated. The dimensions of surface tension are force/length. Consequently, the product of surface tension σ and area has the dimensions of work. Thus $\delta w = \sigma \, dA$ for a differential amount of work, and we can write $W = \int_{A_1}^{A_2} \sigma \, dA$, which becomes $\sigma(A_1 - A_1)$ when σ remains constant. For liquid water at room temperature, σ has the value 73 dynes/cm. How much work would be required to atomize 1 L of water into droplets 10 micrometers in diameter? How much would that same amount of work raise the temperature by stirring? How high could it raise the liter of water?

8 Two Laws From One Dilemma

The caloric theory held that heat was a conserved, weightless fluid. Using this model, Carnot analyzed heat engine cycles and arrived at conclusions that by the mid-nineteenth century had become widely accepted. Meanwhile, Joule's careful experiments had made untenable the conserved-fluid model of heat. Thus, at mid-nineteenth century there was a crisis in heat theory. It did not seem possible that Carnot and Joule could both be right. Nor could either one be wrong. In this chapter, we will find out how and by whom this dilemma was resolved to found modern thermodynamics.

AT THE SAME TIME IN DIFFERENT COUNTRIES

In 1847 the British physicist William Thomson met James Joule at a meeting in Oxford. The son of James Thomson, Professor of Mathematics at Glasgow, William had just the year before accepted the chair in natural philosophy in the same institution. (Because of his great contributions to the development of transatlantic telegraphy, he was knighted in 1866. In 1892 he became Baron Kelvin of Largs and is now more often referred to as Lord Kelvin than as William Thomson. He retired from his professorship in 1899.) William Thomson was a very versatile scholar, publishing over 300 papers that contributed to practically all branches of physical science. However, much of his fame rests upon the work he did as a consequence of his encounter with Joule. Although reluctant to abandon the caloric theory, he finally grasped the implications of Joule's work.

In 1851 he presented a paper on the dynamical theory of heat to the Royal Society of Edinburgh. This paper reconciled Carnot's work with Joule's and brought into perspective the results of the cannon-boring by Rumford, the ice-rubbing by Davy, and pertinent experiments by many others whose efforts we have not mentioned. He set forth the principle of Conservation of Energy, soon universally accepted as the First Law of Thermodynamics. In this same paper, he also proposed the principle of energy dissipation that is one embodiment of the Second Law of Thermodynamics. You will remember that a quarter of a century earlier Sadi Carnot had appealed to another embodiment of the Second Law, the statement that heat did not flow spontaneously from a cold to a hot body. Even Charlie knew that reality, but he didn't call it a law.

It was really Rudolf Clausius who provided the most lucid and elegant early formulation of the first two laws of thermodynamics. William Thomson's contemporary, Clausius was born two years earlier in 1822 at Köslin, Pomerania. He studied at Berlin University from 1840 to 1844, took a degree at Halle near Leipzig in 1848. In 1850 he was appointed Professor at the Royal Artillery and Engineering School in Berlin and privatdocent at the University. (A privatdocent is unsalaried and collects his remuneration directly from the students!) In 1855 he became ordinary professor at Zurich Polytechnic Institute and professor at the University there. (In this context "ordinary" means highest rank, much more prestigious than "extraordinary"!) Twelve years later he moved to Wurzburg as professor of physics and two years after that, in 1869, accepted a similar appointment at Bonn where he remained until he died in 1888. His mobility was in strong contrast to Thomson who stayed sixty years in the same place. But in spite of all the moving around, it took Clausius almost exactly the same amount of time as Thomson to arrive at a clear conception of thermodynamic principles. Thomson was 27 when he presented his paper on the conservation and dissipation of energy in 1851. Clausius was 28 when he read to the Berlin Academy in 1850 his paper "On the Motive Power of Heat and the Laws Which Can Be Deduced from It for the Theory of Heat."

It was probably Clausius who first recognized that there were two principles involved in explaining the conversion of heat to work. One was the equivalence of heat and work that Joule's experiments had placed beyond doubt. (This principle leads to the first law, which embodies the idea of energy conservation as we showed in the last chapter.) Carnot set forth the second principle and recognized that there *must* be heat rejection to a cold reservoir in any continuous process of converting heat from a hot source into work. The second principle also provided that the efficiency of conversion was directly proportional to the difference in temperature between source and sink—that the greater the temperature difference, the greater the amount of work obtainable per unit of heat withdrawn from the source. Carnot had also assumed that the amount of heat rejected to the sink was equal to the amount absorbed from the source, that, as in the case of water falling through a water wheel, it was the *flow* of a conserved quantity of heat from high temperature to low temperature that performed the work.

In those days engines were so inefficient that almost as much heat was discharged to the condenser as was absorbed by the boiler from the firebox. It is not surprising, therefore, that Carnot's assumption of conserved heat flow remained unquestioned. Joule's careful measurements under controlled laboratory conditions eliminated this component of Carnot's theorem. Clausius recognized that Carnot's important conclusions were not undermined by admitting that heat was not a conserved fluid. Indeed, he realized that Joule's doctrine of the conservation of energy and Carnot's principle that efficiency depended upon temperature difference were not only compatible but mutually reinforcing. Carnot's conclusions stemmed from his acceptance of the "obvious" truth that heat could not flow spontaneously from a cooler to a warmer body. We have already pointed out that assertion of this truth is a way of stating what we now call the Second Law of Thermodynamics. As we noted earlier, it is one of history's anachronisms that the Second Law in this form was enunciated 25 years before the First.

A NEW LOOK AT SOME OLD EXPERIENCE

Based on what we have already learned, let's attempt to resolve the Carnot–Joule dilemma. As a first step we apply the First Law to an analysis of some simple heat interactions and consider the consequences. Recall that in Chapter 7 we arrived at the following symbolic statement of the First Law:

$$dE = \delta q - \delta w \tag{1}$$

where dE is a small change in the total energy of a system, and δq and δw are the associated heat and work interactions between the system and its surroundings. This statement asserts the existence of the property energy and defines changes in it as being numerically equal to the difference between the heat absorbed by the system and the work done by it.

Let's apply this relation to the straightforward process of heating a given quantity of some simple substance, for example, a cylinder of air, a bucket of water, or a block of iron. We assume that our system is at rest and does not change position during the heating. Therefore, its kinetic and potential energies do not change and the only possible work interaction relates to a change in volume associated with the ambient pressure, namely, $p\,dV$ work. Under these conditions, the First Law as written in Equation (1) becomes, with some rearrangement of terms:

$$\delta q = dU + p\,dV \tag{2}$$

where dU represents a small change in internal energy, the only kind of energy change allowed under the constraints we have imposed. This equation states that in a small excursion the heat absorbed by the system is numerically equal to the change in internal energy plus the amount of expansion work done by the system.

Now suppose that we confine the system, so that the volume cannot change. Then the last term in Equation (2) disappears and we are left with:

$$\delta q = dU \tag{3}$$

Everybody knows that when a system is heated, its temperature usually increases. In fact, we *defined* the heat capacity of the system as the amount of heat interaction divided by the temperature change that it causes. In symbolic terms:

$$C_V \equiv \left(\frac{\delta q}{dT} \right)_V = \left(\frac{\partial U}{\partial T} \right)_V \tag{4}$$

The subscript V's indicate that the volume remains unchanged, which is the case we are considering. If we happen to be talking about a unit mass of material, for example, a gram or pound, we usually use a lower case c and call c_v the *constant-volume specific heat capacity*. If we have 1 mol of material, we use an upper case C and call C_V the *constant-volume molar heat capacity*.

We have slipped in another bit of symbol convention. When a change in a dependent variable (internal energy, for example) can depend upon changes in more than one independent variable (say, volume and temperature), it is convenient to limit any single term to expressing the amount of change in the dependent variable due to the change in only one of the independent variables, the others being held constant. (See Figure 8-1.) Thus, $\left(\frac{\partial U}{\partial T} \right)_V$ is the ratio of a differential change in U to a differential change in T, V being held constant. Using ∂ instead of d to express the change means that some variables are being held constant. The subscript, V in this case, indicates which variable is the constant one. ∂U and ∂T are called *partial differentials*. The entire term $\left(\frac{\partial U}{\partial T} \right)_V$ is to be read: "the *partial derivative* of U with respect to T, V constant." It is to be taken as the limiting value of the indicated ratio as the denominator (the change in the independent variable) approaches zero. For the case of interactions as opposed to properties, we do not make use of this convention. Thus, we write $\left(\frac{\delta q}{dT} \right)_V$, not $\left(\frac{\partial q}{\partial T} \right)_V$. (Be

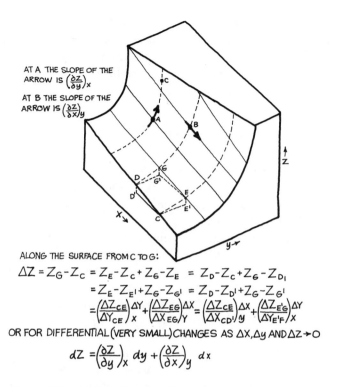

AT A THE SLOPE OF THE ARROW IS $\left(\dfrac{\partial Z}{\partial y}\right)_x$

AT B THE SLOPE OF THE ARROW IS $\left(\dfrac{\partial Z}{\partial x}\right)_y$

ALONG THE SURFACE FROM C TO G:

$$\Delta Z = Z_G - Z_C = Z_E - Z_C + Z_G - Z_E = Z_D - Z_C + Z_G - Z_{DI}$$

$$= Z_E - Z_{EI} + Z_G - Z_{GI} = Z_D - Z_{DI} + Z_G - Z_{GI}$$

$$= \left(\frac{\Delta Z_{CE}}{\Delta Y_{CE}}\right)_x \Delta Y + \left(\frac{\Delta Z_{EG}}{\Delta X_{EG}}\right)_Y \Delta X = \left(\frac{\Delta Z_{CE}}{\Delta X_{CDI}}\right)_y \Delta X + \left(\frac{\Delta Z_{EG}}{\Delta Y_{EIF}}\right)_x \Delta Y$$

OR FOR DIFFERENTIAL (VERY SMALL) CHANGES AS $\Delta X, \Delta y$ AND $\Delta Z \to 0$

$$dZ = \left(\frac{\partial Z}{\partial y}\right)_x dy + \left(\frac{\partial Z}{\partial x}\right)_y dx$$

Figure 8-1 Relations between dependents. Here the value of dependent variable z is determined by the values of x and y. Along any dashed surface line, x is constant; so we can express the variation of z with y by the *partial derivative* $(\partial z/\partial y)_x$, which is the slope of the surface in the y direction along a constant x contour, the dashed line. Similarly, along a solid surface line, y remains constant; and we can write $(\partial z/\partial x)_y$, which represents the rate of change of z with respect to x while y remains constant. Using ∂z and ∂x instead of dz and dx indicates that the derivative is partial in that only *one* of the independent variables is changing (in this case x and in the former case y). We indicate which variable or variables remain constant by the subscripts outside the parentheses. A *total*, as opposed to partial, differential (very small) change in a dependent variable like z can always be represented as a sum of the partial changes. Each partial change is the product of a partial derivative and the differential change in a single variable. For example, in this case, $dz = (\partial z/\partial y)_x dy + (\partial z/\partial x)_y dx$

reminded once again that because q is an interaction, something that happens to the system, we use the prefix symbol δ to indicate a small amount of such interaction. T, like v, p, and U, is a property. Therefore, we use the prefix symbol d, or ∂, to indicate a small change in the property.) In terms of the definition of heat capacity in Equation (4) we can write:

$$\delta q = \left(\frac{\delta q}{dT} \right)_v dT = c_v\, dT \tag{5}$$

This equation is really an old friend. It simply states that the amount of heat interaction between a system and its surroundings is proportional to the change in temperature. The proportionality constant is the constant-volume heat capacity, as before—lower case c_v for a unit mass of substance, upper case C_V for a mol. Indeed, you will recall from Chapter 5 that the unit of measurement for heat interaction is based on the change in temperature of a prescribed amount of a standard reference substance. For example, the calorie was originally defined as the amount of heat interaction required to raise the temperature of 1 g of water by 1°C. Thus, if δq is expressed in calories and dT in degrees C (which are the same size, of course, as Kelvins), then c_v for 1 g of water has the numerical value of unity. Similarly, if δq is in BTU's and dT in degrees Fahrenheit, c_v will be unity for 1 lb of water.

Strictly speaking, these statements are not quite accurate because the calorie and the BTU are defined in terms of heating at constant pressure. As we will learn shortly, there can be an appreciable difference between the constant-volume and constant-pressure specific heats, but it happens to be very small in the case of liquid water. Meanwhile, let's note that we can combine Equations (3) and (5) to obtain for a mole of substance:

$$\delta q = dU = C_V dT \tag{6}$$

which says that in a constant-volume heating process the amount of heat interaction is equal to the change in internal energy that in turn equals the constant-volume molar capacity multiplied by the change in temperature. If the heat interaction is large so that the change in temperature is substantial, we can integrate Equation (6) to obtain:

$$\int \delta q = \int dU = \int_{T_1}^{T_2} C_V\, dT$$

or

$$Q = \Delta U = U_2 - U_1 = C_V(T_2 - T_1) \tag{7}$$

By logging fuel and change in T
He reckons heat capacity.

The last term is correct only if C_V does not change with temperature. If it did, we would have to use an average value of C_V to represent the overall change in U in terms of the change in T.

We have arrived at Equation (7) by assuming that $\delta w = 0$. However, regardless of the kind of interaction that changes internal energy at constant volume, we can continue to equate that energy change with the change in temperature multiplied by the constant-volume heat capacity. In other words, for a process at constant volume, we can *always* write (recalling that upper case letters refer to moles, lower case to unit mass):

$$c_v = \left(\frac{\partial u}{\partial T} \right)_v \quad \text{or} \quad C_V = \left(\frac{\partial U}{\partial T} \right)_v \tag{8}$$

Again: Equation (7) is *always* valid for a process at constant volume. Thus, for an adiabatic process ($\delta q = 0$) at constant volume, such as Joule's whale-oil experiment, we can write for one mole:

$$- \delta w = dU = C_V\, dT \tag{9}$$

where $-\delta w$ is the work done *on* the system by stirring (w is positive when done *by* the system). All of this says that the quantity U, the internal energy, can be changed either by a heat interaction or by a work interaction. The extent of the change is indicated by the change in the easily measured property we call temperature. Moreover, the change in energy for a given temperature change is the same whether it was brought about by a heat interaction, a work interaction, or both. That is the real meaning of equivalence.

So what's so great about this exercise? It is just a recapitulation of what we went through in our consideration of Joule's experiments. On the face of it, these conclusions seem rather obvious. Even so, it took a long time to arrive at them. Between Charlie the Caveman's primitive observations and Joule's precise experiments were perhaps a million years of human endeavor. Progress was slow in part because of careless observation and interpretation. Inaccurate measurement spawned imprecise thought and vice versa. In honor of precision, let's note again that the statement $dU = C_V \, dT$ can be *precisely* inferred from the simple heating experiment only under the constraints that we have carefully specified: No change in volume, phase, or composition; no change in other kinds of energy. To understand why we must be so careful, let's examine another kind of heating process, one in which the pressure, rather than the volume, remains constant.

HEATING AT CONSTANT PRESSURE

Constant-pressure heating is much more common than constant-volume heating. Any heat interaction of an unconfined system really occurs at constant pressure because the ambient atmosphere is almost infinitely elastic and keeps the pressure on the system constant no matter what the volume change. (There are, of course, relatively slow and minor variations in atmospheric pressure that are reflected in the barometer readings so closely followed by sailors and other weather watchers.) In fact, it is quite difficult to carry out a heating process in which there is absolutely no change in volume because practically all materials, including the containers holding the systems under observation, expand at least a little bit when their temperature is increased. Sometimes, the change in volume is very small and can be neglected. Water, in liquid form, for example, has little volume change when heated. There is very little difference between its heat capacities at constant volume and at constant pressure. Air, on the other hand, expands substantially if it is heated at constant pressure. We might expect, therefore, that C_V for air would be quite different from C_p, the constant-pressure molar heat capacity. These intuitive conclusions will be confirmed by the more quantitative consideration that we will now undertake with the help of the First Law.

Let's look at what happens when we heat 1 mole of ideal gas at constant pressure. As before, we assume no change in velocity or position. Therefore, we are concerned only with changes in internal energy U and with work done by a volume change. Thus, we employ the First Law as specified in Equation (2):

$$\delta q = dU + p \, dV \tag{10}$$

Here heat δq absorbed by the system during a small excursion, must provide not only for the change in the system's energy dU, but also for any work that might be done, $p \, dV$.

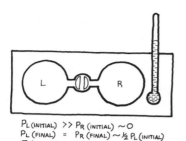

$P_{L\ (INITIAL)} >> P_{R\ (INITIAL)} \sim 0$
$P_{L\ (FINAL)} = P_{R\ (FINAL)} \sim \frac{1}{2}\,P_{L\ (INITIAL)}$
$T_{(INITIAL)} = T_{(FINAL)} = T_{(BOTH)}$

Figure 8-2 Joule's free expansion experiment. Because no temperature change was observed in the bath, there could have been no heat exchange with the gas. Nothing else happened outside the container; so there was no work interaction. We conclude that there could have been no energy change in the gas system. Thus, within the accuracy of the experiment the internal energy of gases like air at modest pressures and temperatures does not depend upon pressure or volume but only upon temperature.

Let's reflect on the term dU in Equation (10). In the case of constant-volume heating, we saw that dU was equal to $C_V\,dT$ because the $p\,dV$ term vanished. But in constant-pressure heating as described by Equation (10), the volume changes by an amount dV. We know there is a change in U associated with any temperature change. The question is whether there is any additional change in U associated with a volume change.

As we have noted, Joule found that when he allowed a gas to expand freely without doing any work, the temperature of the gas did not change. His experiment is represented in Figure 8-2. Two flasks communicating through a valve were immersed in a water bath. Gas on the left-hand side was at high pressure and on the right-hand side at zero pressure, or as near to zero as possible. The valve was opened, so that gas flowed from left to right until the pressure was the same on both sides. The key observation was that there was no change in the temperature of the water bath. Therefore, there could have been no heat interaction between the flasks and the bath. Because the walls of the flask were rigid, there was no displacement outside the system comprising the gas in the flasks. Consequently, no work was done. Therefore, in accordance with Equation (1), because δq and δw were both zero, there could have been no change in total energy. Because there was no change in velocity or position, a zero change in total energy E means also a zero change in internal energy U. But clearly the pressure and the volume of the gas changed. Numerous experiments at various initial conditions of pressure, volume, and temperature always gave the same result: The temperature and the internal energy remained unchanged after the expansion. In sum, there can be a wide range of pressures and volumes associated with a single value of energy. The obvious conclusion is that internal energy U does not depend upon pressure or volume but only upon temperature.

Joule clearly recognized a possible flaw in the experiment. The heat capacity of the gas is quite small relative to the heat capacity of the containers and the water bath. Thus, a fairly large change in gas temperature would show up as a much smaller change in the bath temperature. Although Joule's thermometers could detect temperature changes as small as 0.005°F, he realized that to assign a zero change in temperature to the gas was presumptuous. In conjunction with Kelvin, he later performed an experiment that circumvented the problem of small heat-capacity in the gas. The famous Joule–Kelvin (sometimes called Joule–Thomson) Expansion confirmed the apparent result of the experiment just described—that for most gases under ordinary conditions of temperature and pressure, the internal energy U depends almost entirely upon temperature. As the pressure approaches zero, the dependence upon temperature becomes complete. Moreover, with the formalism developed by Kelvin and Clausius, it became possible to show that this conclusion is rigorously true for any gas obeying the ideal gas equation of state, $pV = RT$. Therefore, we will assume in all our present deliberations that any change in the internal energy of an ideal gas is *completely* reflected in its temperature change. Keep in mind that this assumption is strictly true only if the behavior of the gas is accurately described by $pV = RT$.

The net result of this rather lengthy by-the-way is that we can replace dU by $C_V\,dT$ in Equation (10) *even though that equation relates to a process that does not occur at constant volume.* In other words, because for an ideal gas the internal energy U depends *only* upon temperature, the change in U for *any* process can be represented by some factor multiplied by the temperature change. This factor is precisely the constant-*volume* heat capacity, C_V. Therefore, for 1 mol of ideal gas, Equation (10) becomes:

$$\delta q = C_V\,dT + p\,dV \tag{11}$$

Now recall that, for 1 mol of an ideal gas, $pV = RT$. Thus, if p remains constant, we can write:

$$dp\,V = p\,dV = d\,RT = R\,dT \tag{12}$$

Substitution in Equation (11) gives:

$$\begin{aligned} \delta q &= C_V\,dT + R\,dT \\ &= (C_V + R)\,dT \end{aligned} \tag{13}$$

We define the constant-pressure molar heat capacity C_p as $\left(\dfrac{\delta q}{dT}\right)_p$ so:

$$C_p \equiv \left(\frac{\delta q}{dT}\right)_p = C_V + R \tag{14}$$

Or, the constant-pressure molar heat capacity for any gas obeying $pV = RT$ is larger than the constant-volume molar heat capacity by R, the universal gas constant whose value is 8.314 J/g mol K, or 1.986 cal/g mol K. This very useful relationship is true only for gases and *exactly* true only for gases obeying $pV = RT$. However, it is a pretty good approximation for many gases even when the departure from $pV = RT$ is appreciable. Clearly, for solids and liquids, the difference between C_p and C_V should be much smaller than R because the dependence of their volume upon temperature is very small. They don't expand very much with temperature. Consequently, when heated, they don't do much work against the ambient pressure; and the $p\,dV$ term in Equation (11) is very small. In the case of liquid water, for example, $C_p - C_V$ is only about 0.0023 J/g mol K, or about 0.003 percent. Thus, we were able to assume earlier that the difference between constant-pressure and constant-volume heat capacities for liquid water was negligible.

Before taking up another topic, let's consider Equation (10) again:

$$\delta q = dU + p\,dV$$

When the pressure is constant we can write:

$$\begin{aligned}\delta q &= dU + d\,pV \\ &= d(U + pV)\end{aligned} \tag{15}$$

The sum of U and pV in the last term is encountered so often and is so useful that we give it a name and a symbol of its own. By definition:

$$H \equiv U + pV$$

where H is known as the **enthalpy** of a system (accent on the second syllable). Enthalpy H plays the same role in a constant-pressure process that internal energy U plays in a constant-volume process. For constant-volume heating, $C_V \equiv (\delta q/dT)_V = (dU/dT)_V$; and for constant-pressure heating,

$C_p \equiv (\delta q/dT)_p = (dH/dT)_p$. These relations hold for any substance. In the case of an *ideal gas:*

$$dU = C_V\, dT \tag{16}$$

and

$$dH = C_p\, dT \tag{17}$$

for *any* process, whether or not volume or pressure is constant.

It is instructive to examine the difference between constant-pressure and constant-volume heating of an ideal gas in terms of a pV diagram. Figure 8-3 shows two isotherms. Clearly, T_1 is greater than T_2. We start with gas at point a at T_2, and, by heating at constant volume, we proceed along the **isochore** (constant-volume path) to point b on T_1. We could also reach T_1 by a constant-pressure heating process along the **isobar** (constant-pressure path) to point c. Because the internal energy U of an ideal gas depends only on the temperature, $U_b = U_c$ and $U_b - U_a = U_c - U_a$. Thus, the change in internal energy is the same along either path. Moreover, because $pV = RT$ on either isotherm, $U_b + p_b V_b = U_c + p_c V_c = H_b = H_c$; so the change in enthalpy is also the same along either path. But along the constant-pressure path, the gas did work in an amount equivalent to the shaded area under path ac. The amount of heat required to provide the energy to perform this work makes the constant-pressure specific heat capacity larger than the constant-volume specific heat capacity. No work at all is performed along the constant-volume path, of course.

Note there are any number of possible paths from point a on the T_2 isotherm to the T_1 isotherm. Some possibilities are indicated by the dashed lines. Along any path to the right of path ab, some work will be performed by

Figure 8-3 Several ways to make hot air. There are many possible paths for increasing the temperature of a gas. The most common are adiabatic (work without heat, *ad*), constant volume (heat without work, *ab*), and constant pressure (simultaneous heat and work interactions, *ac*).

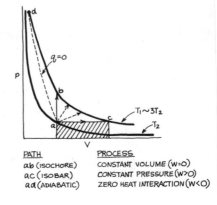

PATH	PROCESS
ab (ISOCHORE)	CONSTANT VOLUME ($W=0$)
ac (ISOBAR)	CONSTANT PRESSURE ($W>0$)
ad (ADIABATIC)	ZERO HEAT INTERACTION ($W<0$)

the gas, and the heat required to raise the temperature to T_1 will be greater than for path ab by an amount equal to the work that is done (the area under the path curve). On all paths to the left of ab, there will be net work done *on* the gas. This work adds energy to the gas and, therefore, reduces the heat required to raise the temperature to T_1. Of particular interest is the adiabatic path along which there is no heat interaction at all. All of the energy required to raise the temperature comes from work. This path is identified by the $q = 0$ legend on the figure. Its characteristics will be considered next.

WORK WITHOUT HEAT—ADIABATIC EXPANSION AND COMPRESSION

We are now ready to carry out a very important calculation with the aid of Equation (11), which represents the First Law for 1 mol of gas that doesn't change its velocity, position, phase, or composition. We shall be writing it again and again to engrave it on your memory:

$$\delta q = C_V \, dT + p \, dV$$

Bear in mind that this statement is strictly true only for a gas obeying $pV = RT$ and performing no work except by expansion. Let's consider what happens if we carry out an adiabatic expansion or compression, that is, one without any heat interaction. For such a process, δq is zero. From Equation (11) we can write:

$$\delta w = p \, dV = -C_V \, dT \tag{18}$$

This resembles Equation (9), but there we demanded constant volume. Now we have removed that constraint because $dU = C_V \, dT$ for an ideal gas even when volume is not constant. For a substantial excursion, or change, we can write (assuming C_V is constant by integrating Equation (18)):

$$W = \int_{V_1}^{V_2} p \, dV = -C_V(T_2 - T_1) \tag{19}$$

Note that to obtain the amount of work done we don't have to work out the summation represented by $\int_{V_1}^{V_2} p \, dV$. That calculation would be a bit difficult because p changes during the process. All we must know is the initial and final temperature and the constant-volume heat capacity. In words, the amount of work done during an adiabatic expansion is determined entirely by the temperature change and the heat capacity.

We'll use this information shortly, but, for the moment, let's manipulate Equation (18). We can replace p by RT/V as we have done before:

$$\frac{RT}{V}\, dV = -C_V\, dT \tag{20}$$

We cannot integrate this equation without further information because T varies with V. But we can divide both sides by T to obtain:

$$\frac{R}{V}\, dV = -\frac{C_V\, dT}{T} \tag{21}$$

Both sides of this equation are very much like the expression we integrated in Chapter 4 to determine the work of isothermal expansion:

$$\delta w = p\, dV = RT\frac{dV}{V}$$

You will remember, or if you don't you can look back and find out, that when we integrated this expression over a substantial change in volume we found:

$$W = RT \ln \frac{V_2}{V_1}$$

(where ln means the natural logarithm of). We obtained this result because R and T were both constant. The left-hand side of Equation (21) is essentially the same form, dV/V preceded by a constant R. Similarly, on the right-hand side we have dT/T preceded by a constant C_V. Thus, integrating both sides of this equation gives:

$$R \ln \frac{V_2}{V_1} = -C_V \ln \frac{T_2}{T_1} \tag{22}$$

The important conclusion to draw from these calculations is that for an adiabatic process in an ideal gas, the ratio of the final volume to the initial volume is determined entirely by the ratio of the final temperature to the initial temperature, no matter what the pressure (we'll use this information later on). Some of you may remember that $n \ln X = \ln X^n$. This relation lets us divide both sides of Equation (22) by C_V to obtain:

$$\ln \left(\frac{V_2}{V_1} \right)^{R/C_V} = -\ln \frac{T_2}{T_1} = \ln \frac{T_1}{T_2} \tag{23}$$

If we take the antilogarithm of both sides,

$$\left(\frac{V_2}{V_1} \right)^{R/C_V} = \frac{T_1}{T_2} \tag{24}$$

A little more manipulation gives:

$$T_1 V_1^{R/C_V} = T_2 V_2^{R/C_V} \tag{25}$$

Because our choice of states 1 and 2 was perfectly arbitrary, the relation of Equation (25) will clearly hold for any pair of states during an adiabatic process. Thus, we can write the well-known relation for an adiabatic process in an ideal gas:

$$TV^{R/C_V} = \text{constant} \tag{26}$$

If you manipulate further by using the ideal gas law $pV = RT$, the result of Equation (17) that $C_p - C_V = R$, and the definition that $\gamma = C_p/C_V$, you will find that Equation (26) will lead to:

$$TV^{\gamma-1} = \text{constant} \tag{27}$$

$$pV^{\gamma} = \text{constant} \tag{28}$$

$$Tp^{(1-\gamma)/\gamma} = \text{constant} \tag{29}$$

These three equations give us the relations between gas properties during an adiabatic process in an ideal gas.

We have developed a number of relations characterizing the behavior of an ideal gas during various kinds of processes. We summarize them in Table 8-1. Note that $pV = RT$ is always true for an ideal gas and that the first three relations are implicit in it. The last three adiabatic relations cannot be derived from $pV = RT$ alone, but require invocation of the First Law as well. Also implicit in every case, though we have not spelled it out, is that the process takes place slowly enough that the gas is uniform and can be

TABLE 8-1 pVT relations for an ideal gas in processes keeping something constant.

Process	Relation
Constant temperature ($dT = 0$)	$pV = $ constant
Constant pressure ($dP = 0$)	$V/T = $ constant
Constant volume ($dV = 0$)	$p/T = $ constant
Adiabatic ($q = 0$)	$pV^\gamma = $ constant
Adiabatic	$TV^{\gamma-1} = $ constant
Adiabatic	$Tp^{1-\gamma/\gamma} = $ constant

characterized by a temperature and a pressure at all times. Otherwise, we could not presume $pV = RT$. This condition of uniformity usually can occur only when the expansion is relatively slow because it is resisted and, therefore, does work. It is equivalent to the condition of reversibility considered earlier.

CARNOT'S CYCLE REVISITED

We are now ready to reexamine Carnot's cycle in light of the First Law and to complete the analysis we began in Chapter 6. You may remember that this cycle consists in subjecting a gas to the following processes in succession: (1) an isothermal ($dT = 0$) expansion at a high temperature T_1; (2) an adiabatic ($\delta q = 0$) expansion from T_1 to a lower temperature T_2; (3) an isothermal compression at T_2; and (4) an adiabatic compression from T_2 back to T_1. Once again we show this cycle on a pV diagram (Figure 8-4). We assume that our system is a 1 mol of ideal gas whose equation of state is $pV = RT$.

We start at point a and allow the gas to expand along the isotherm from a to b. During this expansion it is in contact with a reservoir at T_1 that maintains the gas temperature constant at T_1. For this process we can write the First Law as obtained from an integration of Equation (10):

$$Q_{ab} = (U_b - U_a) + W_{ab}$$
$$= C_V(T_b - T_a) + \int p\, dV$$
$$= 0 + \int_{V_a}^{V_b} RT_1 \frac{dV}{V}$$
$$= RT_1 \ln \frac{V_b}{V_a} = W_{ab} \qquad (30)$$

The final expression for the work done *by* the gas during an isothermal

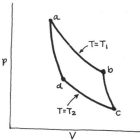

CLOCKWISE: HEAT ABSORBED $a \rightarrow b$
HEAT REJECTED $c \rightarrow d$
WORK PRODUCED
COUNTERCLOCKWISE: HEAT ABSORBED $d \rightarrow c$
HEAT REJECTED $b \rightarrow a$
WORK REQUIRED

Figure 8-4 Either way it's still a Carnot cycle.

expansion is just as we calculated it in Chapter 4. Note that because the temperature is constant, the internal energy does not change. Therefore, the work done by the gas equals the heat it absorbs.

The next step is the adiabatic process from b to c, during which the gas is insulated from any heat interaction and the temperature decreases to T_2. For this step, the First Law is:

$$Q_{bc} = (U_c - U_b) + W_{bc}$$
$$0 = C_V(T_c - T_b) + W_{bc}$$
$$W_{bc} = -C_V(T_c - T_b) = C_V(T_1 - T_2) \tag{31}$$

Thus, the work done *by* the gas during the adiabatic expansion is the product of the constant-volume specific heat C_V and the change in temperature $(T_1 - T_2)$. It was this adiabatic work that we could not calculate when we first considered the Carnot cycle. Carnot himself was in the same boat. He did not have the First Law either.

The next leg of the cycle is the isothermal compression from c to d. The result is precisely analogous to Equation (30):

$$Q_{cd} = RT_2 \ln \frac{V_d}{V_c} = W_{cd} \tag{32}$$

Note that W_{cd} is again the work done *by* the system. In this case, it is a negative quantity because the system volume decreases; that is, the surroundings are doing work on the system. The signs are all taken care of in the computation because V_d is smaller than V_c, and the logarithm of a number less than unity, V_d/V_c, is always negative.

The final leg of the cycle is an adiabatic compression from d to a. In analogy with Equation (31), we can write:

$$
\begin{aligned}
W_{da} &= -C_V(T_a - T_d) \\
&= -C_V(T_1 - T_2)
\end{aligned}
\tag{33}
$$

Again, the work done *by* the system is a negative quantity because the system is compressed by (i.e., worked on by) the surroundings. The right-hand term in Equation (33) reflects this negative character: $(T_1 - T_2)$ is positive as is C_V; so the negative sign persists.

Clearly, the total net work done by the system during the cycle will be the sum of the work terms for each step:

$$
W_{net} = W_{ab} + W_{bc} + W_{cd} + W_{da}
\tag{34}
$$

From Equations (31) and (33), we note that the work terms in the adiabatic legs W_{bc} and W_{da} are equal in magnitude and opposite in sign. Therefore, their sum is zero. Consequently, the net work is simply the sum of the two isothermal steps, W_{ab} and W_{cd}. (Note that we now have proved what we were forced to assume in Chapter 6.) By referring to Equations (30), (32), and (34), we can write:

$$
W_{net} = RT_1 \ln \frac{V_b}{V_a} + RT_2 \ln \frac{V_d}{V_c}
\tag{35}
$$

What we are usually most interested in is the ratio of work produced (what we get from the engine) to the amount of heat withdrawn from the high-temperature source (what it costs us in terms of fuel). This ratio is what we have called the efficiency. We can see from Equation (30) that the heat withdrawn from the high-temperature reservoir, Q_{ab}, is the same as the work done during the isothermal expansion at T_1, $RT_1 \ln(V_b/V_a)$. Thus, for efficiency:

$$
\frac{W_{net}}{Q_1} = \frac{RT_1 \ln(V_b/V_a) + RT_2 \ln(V_d/V_c)}{RT_1 \ln(V_b/V_a)}
\tag{36}
$$

where we have written Q_{ab} as Q_1 to emphasize that it is the heat absorbed from the reservoir at T_1.

Let's now return to Equation (22), arrived at during our consideration of adiabatic processes earlier in this chapter:

$$R \ln(V_2/V_1) = -C_V \ln(T_2/T_1) = C_V \ln(T_1/T_2) \tag{22}$$

This equation says that the ratio of initial and final volumes for an ideal gas in adiabatic compression or expansion is determined solely by the initial and final temperatures. In the cycle we have been considering, the initial and final temperatures in both adiabatic legs are either T_1 or T_2. Therefore, we can write:

(a) For leg bc: $C_V \ln(T_2/T_1) = R \ln(V_b/V_c)$
(b) For leg da: $C_V \ln(T_2/T_1) = R \ln(V_a/V_d)$ $\qquad$ (37)

from which we get:

$$V_b/V_c = V_a/V_d \tag{38}$$

A little further manipulation yields:

$$V_b/V_a = V_c/V_d \tag{39}$$

Therefore, because $\ln V_a/V_c = -\ln V_c/V_a$, if you cancel like terms in the numerator and denominator of Equation (36) you will finally obtain:

$$\frac{W}{Q_1} = \frac{T_1 - T_2}{T_1} \tag{40}$$

This simple but famous relation asserts that the ratio of the amount of work obtained from an engine operating between a high-temperature reservoir and a low-temperature reservoir, to the amount of heat absorbed from the high-temperature reservoir, is equal to the ratio of the temperature difference between the two reservoirs to the temperature of the high-temperature reservoir. (Note the economy of mathematical expression. That last long sentence used 53 words to say what Equation (40) says in 9 symbols.) Though the final expression is simple and terse, the argument that led to it has been long and perhaps tedious. Even if you are unwilling or unable to follow all the steps, you can take as an article of faith the validity of Equation (40). The logic has been examined so many times by so many people and the consequences have so often survived experimental test that there can be no doubt that this famous equation embodies fundamental truth about the real world. Indeed, its ingredients include the Zeroth Law (existence of temperature) and the First Law (conservation of energy). That it relates also to the Second Law we shall see shortly.

In addition to the two laws, there have been some other important assumptions and limitations underlying our development of Equation (40). We have assumed that the working fluid is an ideal gas that obeys the equation of state, $pV = RT$. (Remember that the temperatures in Equation (40) are those for which this equation of state holds, namely, the ideal gas thermometer scale or the absolute scale, Kelvin or Rankine temperatures. You cannot use just any temperature and expect the expression to be meaningful.) We have also assumed that each step takes place *reversibly*. This limitation means there has been no mechanical friction and that all the heat exchanges have occurred at infinitesimal temperature differences and, therefore, at infinitesimally slow rates. It also means that the cycle could be run in either direction. In a clockwise direction on a *pv* diagram, work is produced while heat is absorbed from the high-temperature reservoir and rejected to the low-temperature reservoir. In a counterclockwise direction, work is consumed while heat is absorbed at low temperature and rejected at high temperature. Equation (40) holds no matter which direction the cycle follows. Finally, we note that, as derived, Equation (40) is specific to a very particular cycle comprising two isothermal processes and two adiabatic processes.

In view of all these constraints and limitations, it might seem strange that we attach so much significance to Equation (40). But that notion doesn't reckon with Carnot's genius. Although he was unable to complete the calculation of the amount of net work produced by his ideal cycle, he was able, as we showed in Chapter 6, to demonstrate that *any* working fluid would have to provide the same efficiency or else there could be a net flow of heat from a cooler to a warmer body. Moreover, and for the same reason, no other *cycle* could have a higher efficiency, that is, could provide more work per unit of heat absorbed at high temperature. In short, Equation (40) represents the *maximum possible efficiency* for *any* conceivable engine that absorbs heat at T_1 and rejects heat at T_2.

We indicate this generalization by:

$$\frac{W}{Q_1} \leq \frac{T_1 - T_2}{T_1} \tag{41}$$

(where the symbol $\leq$ reads ''equal to or less than''). In this form the relation means that the efficiency of *any* heat engine operating between one reservoir at high temperature and another one at low temperature can never be greater than the ratio of the temperature difference between the two reservoirs to the temperature of the hot one. This assertion is one possible statement of the Second Law of Thermodynamics. As we have shown, it is equivalent to the assertion that heat cannot spontaneously flow from a low temperature to a high temperature. This latter assertion can also be considered as a statement of the Second Law. It was first invoked by Carnot but in this explicit form is usually called the Clausius statement of the Second Law.

A third equivalent assertion known as the Kelvin–Planck statement of the Second Law says that no cyclic process (engine) can have as a *sole* conse-quence the absorption of heat from a reservoir and the performance of work. This statement was also implicit in Carnot's recognition that to perform work continuously an engine must reject heat to a sink as well as absorb heat from a source. In sum, it is fair to say that Equation (40) is an explicit statement of the Second Law that implicitly depends also upon the Zeroth and First Laws. Clearly, it is aphoristic. In the next chapter, we will see that it is also useful.

Before closing we note one useful variation of Equation (40). In deriving it, we chose to evaluate the ratio of work performed to the heat withdrawn from the high-temperature source, or W/Q_1. We could just as well have related the work to the heat rejected to the low-temperature sink, or W/Q_2. If you reexamine the sequence of arguments beginning with Equation (34), you will find by making the appropriate substitutions that:

$$\frac{W}{Q_2} \leqq \frac{T_1 - T_2}{T_2} \tag{42}$$

Symmetry might have led you to expect this relation. Indeed, because of symmetry, it should be easy to remember both Equations (41) and (42). Usually, we are interested in the relation between work and heat interaction with the low-temperature reservoir when we are using work to cool some-thing—when we are running the cycle *counterclockwise* and using it as a refrigerator. In this case, Q_2 is what we "get" and W is what we "pay," so we often find Equation (42) written upside down as:

$$\frac{Q_2}{W} \leqq \frac{T_2}{T_1 - T_2} \tag{43}$$

The ratio Q_2/W is called the **coefficient of performance**, or CEP, of a refrigerator.

Note that the efficiency of a heat engine, W/Q_1, as described in Equation (41) can never be greater than unity. You can prove this to yourself by trying a few combinations of T_1 and T_2. Because T_1 is greater than T_2, the denominator of the right-hand side must always be greater than the numerator. There is no such constraint on the ratio, W/Q_2. Some trials with various combinations of T_1 and T_2 will persuade you that any value, from much less than unity to much greater than unity, can be realized with entirely reasonable choices for T_1 and T_2.

You may be a bit puzzled by the fact that if you invert Equation (43) as it is written, at the same time remembering to reverse the direction of the inequality, you would obtain:

$$\frac{Q_2}{W} \gtreqless \frac{T_2}{T_1 - T_2} \tag{44}$$

This statement is true even though it seems to be the opposite of Equation (42) because the inequality has changed direction. The difference is that in the case of Equations (41) and (44), we are talking about an *engine* going clockwise around the cycle on a pV diagram, withdrawing heat at T_1 and rejecting it at T_2. In the case of Equation (43), we are referring to a *refrigerator* going counterclockwise around the cycle on a pV diagram, withdrawing heat from T_2 and rejecting it to T_1.

By the kind of argument Carnot used to prove maximum efficiency, we can readily show why (44) is true for an engine, and (43) is true for a refrigerator. In the case of an engine, any heat that flowed directly from T_1 to T_2 without going through the engine would increase the value of Q_2 (heat flow to the low-temperature reservoir) without increasing W. Therefore, W/Q_2 would be less than its maximum theoretical value of $(T_1 - T_2)/T_2$; thus, Q_2/W would be greater than $T_2/(T_1 - T_2)$. In the case of the refrigerator (Equation (42)), any *natural* addition to Q_2 (leak from T_1 to T_2) would represent a *decrease* in the net flow of heat Q_2 from T_2 to T_1. Therefore, Q_2/W would be less than its maximum possible value of $T_2/(T_1 - T_2)$. The situation is represented in Figure 8-5. The widths of the shaded channels represent the relative energy flows.

With judicious choice of tool
Work can either heat or cool.

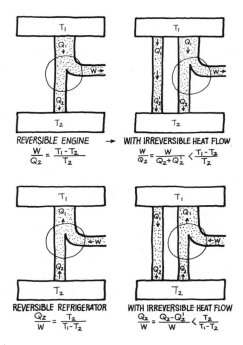

REVERSIBLE ENGINE $\rightarrow$ WITH IRREVERSIBLE HEAT FLOW

$$\frac{W}{Q_2} = \frac{T_1 - T_2}{T_2}$$

$$\frac{W}{Q_2} = \frac{W}{Q_2 + Q_2^!} < \frac{T_1 - T_2}{T_2}$$

REVERSIBLE REFRIGERATOR

$$\frac{Q_z}{W} = \frac{T_2}{T_1 - T_2}$$

WITH IRREVERSIBLE HEAT FLOW

$$\frac{Q_2}{W} = \frac{Q_2 - Q_2^!}{W} < \frac{T_2}{T_1 - T_2}$$

Figure 8-5 Heat leaks waste work because they are irreversible.

You may think that this reasoning is awkward and unwieldy. But if you do some simple trial calculations, you will find that the steps and their sequence in the arguments are quite simple and logical. However simple and rigorous they may be, many people have decried the inelegance of resorting to such artificial constructs as frictionless engines and refrigerators in statements of fundamental physical laws. They favor mathematical constructs that may be more elegant but are much more remote from our empirical experience. Charlie would not like them at all. Even so we will examine some of them in the last two chapters.

EXAMPLE EXERCISE

Suppose an iceberg weighing 10^{10} kg drifts into the Gulf Stream which has a temperature of 22°C. If we operate a heat engine using the Gulf Stream as a "hot" source and the iceberg as a "cold" sink, what is the maximum amount of work that could be generated while the iceberg is melting? Assume that the iceberg's temperature is 0°C.

1. The heat absorbed by melting 1 kg of ice is close to 80 kcal. There are 4186 J/kcal. Therefore, the melting iceberg could absorb $4186 \times 80 \times 10^{10}$ J, or 3.35×10^{15} J.

2. According to the Heat Engine Rule, the maximum amount of work W that can be obtained by an engine exhausting heat Q_2 at T_2 is

$$W = Q_2(T_1 - T_2)/T_2$$

Note that we are given the amount of heat Q_2 rejected to the low-temperature sink at T_2.

3. In this case T_2 is the melting point of ice, or 273 K. The temperature of the source T_1 is 22°C or 295 K. Therefore,

$$W = Q_2(295 - 273)/273 = Q_2 \times 0.08$$

Q_2 is 3.35×10^{15} J. Therefore, W is $0.08 \times 3.35 \times 10^{15}$ or 2.68×10^{14} J. That is equal to about three day's work for a modern, 1000-megawatt power plant.

SUMMING UP

At the end of the preceding chapters we have summarized important points. There are so many important consequences of what we have described in this chapter that we are going to devote the next chapter to a summary of them. We will also include material from previous chapters that has been prerequisite to this discussion.

LAW PRACTICE

1. Ten kg mol of ideal gas are heated from 400 K to 800 K. The initial pressure is 2 atm. Determine the amount of heat interaction required and the final pressure and volume:
   (a) If the heating is at constant volume.
   (b) If the heating is at constant pressure.
   (c) How much work is done by the gas in each case? Assume $C_V = 4R$.

2. Suppose 1 kg mol of ideal gas with $C_p = 5R$ at an initial temperature of 500 K has its volume quartered while being compressed at a constant pressure of 5 atm.

(a) How much heat is rejected?

(b) How much work is required?

(c) If the gas were allowed to return to its original volume while the temperature was maintained at the value reached after compression, how much heat would be required?

3. During the compression stroke of a diesel engine, air is compressed from 1 atm and 27°C to $\frac{1}{25}$ of its initial volume. Assume that C_V for air is $2R$ and that the compression is adiabatic.

(a) How much work is required per gram-mole?

(b) What are the final temperature and pressure?

4. An inventor claims to have developed an engine that takes in 1000 BTU/sec at a temperature of 600 K, rejects heat to a sink at 300 K, and develops 875 horsepower. Recall that 1 horsepower is 550 ft lbf/sec. Would you invest money to put this engine on the market?

5. A Carnot refrigerator is operated between reservoirs at 0°C and 100°C.

(a) If 1000 joules of heat are absorbed from the low-temperature reservoir, how many joules are rejected to the high-temperature reservoir?

(b) What is the coefficient of performance?

6. An iceberg having a mass of 10^8 kg at a uniform temperature of 0°C is floating in ocean water at 10°C.

(a) If a heat engine uses the ocean water as a source and the melting iceberg as a sink, how much work could be produced if you assume that all the heat needed to melt the iceberg is from the engine's heat rejection?

(b) If that work were used to pump heat from the ambient ocean into a boiler at 100°C, how much water could be vaporized in the boiler? Assume that the heat of fusion for ice is 80 cal/g and the heat of vaporization of water is 540 cal/g.

7. The temperature in a household refrigerator is maintained at 0°C, and the temperature of the room where it is located is 25°C. The heat leaking into the refrigerator from the warm room is 8×10^6 J every 24 hours, enough to melt about 50 lbm of ice. This heat must be removed from the refrigerator to maintain its temperature at 0°C.

(a) If an ideal Carnot refrigerator were available, how many watts would be required to operate it?

(b) If electric power costs 10 cents/kWh and ice costs two cents/lbm, how much more or less would it cost to use ice?

8. A 155-mm howitzer with a barrel 2 m long is charged with 1 kg of propellant that occupies 20 cm of the barrel length. The projectile has a mass of 2 kg. When the gun fires, the powder generates gas at 2400 K essentially instantaneously, that is, before the projectile moves. Assume that the molecular weight of the gas is 30, its specific heat at constant volume is $3R$, and the gas expansion is adiabatic.
   (a) What is the muzzle velocity of the projectile?
   (b) If it were fired straight up, how high would it go?

9. The internal energy of a particular gas can be expressed as $U = A + BpV$ where A and B are constants. During adiabatic expansion pV^n remains constant. Find C_V, C_p and the value of n assuming that the gas obeys $pV = RT$. Express your answers in terms of A, B, and R. (Equate expressions for dU obtained from the First Law and from the expression above.)

10. The motor of a household refrigerator consumes electric power at an average rate of 100 W when its internal temperature is 7°C and the room temperature is 27°C. The door switch controlling the light in the refrigerator sticks and the bulb remains on when the door is closed. The power consumption then increases to 105 W. Assume that the refrigerator's overall coefficient of performance is half that of an ideal Carnot machine. What is the wattage of the light bulb? Assume that all the power to the bulb gets dissipated as thermal energy inside the refrigerator.

11. One mol of ideal gas initially at p_o, T_o, and V_o goes through the following cycle: an isothermal expansion from a to b that doubles the volume, a constant volume pressure increase from b to c, and a constant pressure compression back to the initial conditions. Let $C_V = 2R$.
   (a) In terms of R, T_o, and V_o determine p, V, and T at b and c.
   (b) Sketch the cycle on a pV diagram.
   (c) Determine Q and W for each leg and then the efficiency of the cycle assuming a clockwise direction. Express your results in terms of R and T_o.

12. A solar hot-water heater raises the temperature of 200 liters of water from 27°C to 47°C. Suppose a heat engine with 40 percent of the efficiency of a Carnot engine had absorbed the same amount of heat at 600 K and rejected waste heat at 300 K. If the output of that engine were converted into electric power with an efficiency of 90 percent and that electric power were dissipated in a resistance heater, how much water would have been heated through the same temperature interval, that is, 27°C to 47°C?

9　All in a Nutshell

Early in this short history of heat, the pace of advance was pretty slow. We were able to include several centuries of progress per chapter. That it took three chapters to describe progress during the first half of the nineteenth century implies a lot had happened after 1800. It is appropriate, therefore, to summarize the state of thermodynamics after Kelvin and Clausius finished putting it together. We won't express ideas in quite the way they did, but the essentials are the same. Not much new has since been added.

1. Systems, States, and Properties.　That part of the real world that we focus our attention and do our accounting on is called a *system*. The *state* of a system is characterized by a particular set of values for its *properties*— quantities like pressure, volume, mass, velocity, and composition. Property values are simply numbers resulting from carefully prescribed measuring operations. Associated with these numbers are dimensions suggested by the kind of measuring operation. *Mechanical properties,* for example, are those that can be expressed completely in terms of mass, length, and time.

2. Temperature.　The property that measures the hotness of a system is called *temperature*. Thermometers usually indicate temperature by a mechanical property like length or pressure, or an electrical property like resis-

By its temperature he may know
Just how hot is this volcano!

tance. But temperature cannot be expressed completely in terms of mechanical or electrical properties. Its evaluation always depends on some reference hotness like the ice point where water freezes or the steam point where it boils. Thus, temperature adds a new dimension to the description of physical systems. The commonly used Celsius scale assigns the value 0°C to the ice point and 100°C to the steam point. Temperatures on its absolute counterpart, the Kelvin scale, are given by $T(K) = t\,°C + 273.15$. The Fahrenheit scale assigns 32°F to the ice point and 212°F to the steam point. Absolute Fahrenheit temperatures are known as Rankine temperatures and are given by $T(R) = t°F + 459.67$ (upper case T is usually used for absolute scales). The definitive standard reference temperature is now the triple point of water where ice, liquid water, and water vapor coexist. It is assigned the value 273.16 K.

   3. Equation of State.   For every substance there is a relation among its properties called an *equation of state*. The behavior of many gases is approximated by the simple relation $pV = nRT$ where p is pressure, V is volume, T is absolute temperature, n is number of moles (amount of gas), and R is a universal constant. Liquids, solids, and dense gases involve much more complex relations.

   4. Interactions.   The state of a system is changed when one or more of its property values change. If there is a one-to-one correspondence, or correlation, between a state change in a system and a state change in the surroundings or another system, we say that there has been an *interaction* between the system and the surroundings (or the other system). Something has happened across the boundary of the system.

Every fish bite rings the bell.
An interaction sure as—(well?)

   5. Heat.   The interaction that occurs when two systems at different temperatures, T_1 and T_2, are brought together is called *heat*. The symbol Q represents a substantial amount of heat interaction. A small amount is represented by the symbol δq. A heat interaction generally diminishes the temperature difference, $T_1 - T_2 = \Delta T$, that causes it. If $T_1 > T_2$ and if the temperatures change, T_1 will always decrease and T_2 will always increase. If allowed to continue indefinitely, a heat interaction between finite systems will ultimately eliminate the temperature difference that brings it about. Sometimes, the consequence of a heat interaction will be a change in composition rather than a change in T_1 and T_2. Melting or freezing, and vaporizing or condensing are examples of phase changes that can occur. Changes in chemical composition may also maintain a temperature constant during a heat interaction.

6. Insulators.   Some substances placed between two bodies at different temperatures slow down the rate of heat interaction. They are called *thermal insulators*. Cork, fur, feathers, foamed plastics, and evacuated regions (especially with reflective surfaces) are examples of effective insulators.

A bearskin coat for winter wear
Will insulate from frigid air.

7. Measurement of Heat.   A heat interaction Q is measured in units that are defined in terms of a standard, reference heating process. The *calorie* is the amount of heat interaction that raises the temperature of 1 g of water from 14.5°C to 15.5°C. One *British Thermal Unit,* BTU, is the amount of heat interaction necessary to raise the temperature of 1 lb of water from 59.5°F to 60.5°F. Since 1968 values for calories and BTU's has been defined in terms of work units: 1 cal = 4.1860 J; 1 BTU = 778.28 ft lb.

8. Flow of Heat.   The direction of Q is from the higher T to the lower. Because it was long thought that the mechanism of a heat interaction consisted in the transfer of a conserved weightless fluid, caloric, Q is frequently described as a flow of heat from a high-temperature *source* to a low-temperature *sink*. The source is considered to lose or reject heat. The sink is considered to gain or absorb heat. These terms and this imagery are so embedded in the language that they prevail even though the model that spawned them has long since been discredited.

9. Work.   Any interaction between a system and its surroundings that is not heat is called work. The usual symbol for a very small amount of work interaction is δw. Substantial amounts of work interaction are represented by the symbol W. The most elementary empirical test to distinguish between heat and work is to examine the effect of inserting a thermal insulator between the interacting systems. If the interaction slows down, it is heat. If unaffected, it is work.

10. Kinds of Work.   There are many forms of work. All of them involve the product of a force and a displacement. Some frequently encountered combinations are shown in Table 9-1.

TABLE 9-1   Force-displacement combinations for various kinds of work.

Kind of force	Displacement variable	δw
F—simple mechanical	x—linear distance	$F\,dx$
p—pressure (F/area)	V—volume	$p\,dV$
mg—gravitational on mass m	h—height	$mg\,dh$
ϵ—electromotive (voltage)	q—charge	$\epsilon\,dq$
μ—chemical potential	n—quantity of mass (mol)	$\mu\,dn$

11. Interchangeability of Work.   All forms of work are freely interconvertible, that is, they can be changed from one to another, in principle, with no losses. For this reason, work is sometimes defined as any interaction between a system and its surroundings that could be reduced to the raising of a weight as the *sole* effect in the system or the surroundings. If the weight goes up or *could* have gone up in the surroundings, the system is said to perform positive work on the surroundings.

This interaction work must be.
A rising weight is all we see.

12. Equivalence of Heat and Work.   Joule's careful experiments showed that some changes in the state of a system could be equally well brought about by either heat or work interactions. In such cases there was always the same ratio between the amounts of heat and work necessary for a particular change; that is, $W = JQ$ where J is found to be 778 ft lb/BTU, or 4.186 J/cal. The equivalence of heat and work means that either interaction can be measured in any of the traditional units. W can be expressed in calories or BTU's as well as joules or foot-pounds. Similarly, Q can be expressed in ergs or newton-meters as well as in calories or BTU's. When measured in the same units, the equivalence of heat and work for a given amount of change can be expressed as $W = Q$.

13. Energy and the First Law.   The experimentally demonstrated equivalence of heat and work led to a generalization of the principle of *conservation of energy* in mechanical systems to include thermal effects, the *First Law of Thermodynamics*. In words: Any change in the total energy E of a system is described by the difference between the heat Q absorbed by the system and work W done by the system. In symbolic form a small energy change is expressed:

$$dE = \delta q - \delta w$$

For larger changes:

$$\Delta E = Q - W$$

14. Kinds of Energy.   The total energy of a system E can comprise many kinds. Thus, we can write:

$$dE = mg\, dh + m\, d(v^2/2) + dU + \cdots$$

where m is mass, g the acceleration due to gravity, v is velocity, and h is height in the gravitational field. As many additional terms for electrical, chemical, magnetic, and strain effects are added as are needed to describe accurately the system's processes. The three terms shown in the above equation are those with which we are most concerned. In order, they relate to gravitational potential energy, kinetic energy, and *internal energy*. The last of these, dU, is "new" to mechanics and represents a recognition of thermal effects in real mechanical systems.

15. Internal Energy.   Changes in internal energy U manifest themselves as changes in temperature, phase, or chemical composition. They can be effected by either heat or work interactions. It is fair to say that U is the only form of energy that responds directly to heat interactions. When two bodies, A and B, at different temperatures undergo a heat interaction in the absence of work, the hotter body loses internal energy and the cooler one gains it. Symbolically, in terms of the First Law, this process of heat interaction can be represented:

For system A at high temperature:

$$\Delta U_A = U_{A_2} - U_{A_1} = Q_{AB} < 0 \qquad \text{(i.e., negative)}$$

For system B at low temperature:

$$\Delta U_B = U_{B_2} - U_{B_1} = -Q_{AB} = Q_{BA} > 0 \qquad \text{(i.e., positive)}$$

Thus, the heat interaction can be regarded as a transfer of internal energy from A to B. In this sense internal energy U plays the role once assigned to caloric fluid, the flow of which was to account for heat interactions. Theoretically, caloric fluid was conserved, neither created nor destroyed. The internal energy of a system, as Joule showed quantitatively, can be changed by work interactions as well as by heat interactions. Thus, internal energy can in no sense be regarded as a conserved substance. When only heat interactions are involved, internal energy is conserved and the caloric fluid model accurately accounts for what happens. But because of its limited applicability (i.e., to cases in which $W = 0$), it cannot be considered as a complete embodiment of truth.

16. Internal Energy and Constant-Volume Heating.   The intimate and direct relation between Q and U provides a basis for measuring changes in U. Recall that the First Law requires:

$$dE = \delta q - \delta w$$

If we consider simple systems engaging only in $p \, dV$ work—systems that do not change position, velocity, or composition, and do not encounter electrical or magnetic effects, then $dE = dU$ and we can write with some rearrangement:

$$\delta q = dU + p \, dV$$

We ask how T changes with a heat interaction and divide both sides by dT:

$$\delta q/dT = dU/dT + p(dV/dT)$$

First, we consider the case when volume remains constant, so that $dV = 0$. Then for one mole;

$$C_V \equiv (\delta q/dT)_V = (\partial U/\partial T)_V$$

where the identity sign $\equiv$ defines C_V, the *constant-volume molar heat capacity*. Clearly, $dU = C_V\, dT$ for a constant-volume process. (That the volume does not change is indicated by the subscript V.) This equation says that the change in internal energy U for a constant-volume process is numerically identical with the product of the constant-volume specific heat capacity C_V and the temperature change dT. For most liquids and solids during a heating process, ΔV is very small even though the system is not constrained to constant volume. For such systems $dU = C_V\, dT$ is a very good approximation even if the volume is not constant. In the case of gases, if $pV = RT$, then U depends only upon T and $dU = C_V\, dT$ for any process whether or not the volume changes. Because $pV = RT$ is a pretty good approximation for many gases over quite a wide range of conditions, $dU = C_V\, dT$ is also a fairly good approximation for many gaseous processes.

17. Enthalpy and Constant-Pressure Heating.   More common and more important than heating at constant volume is heating at constant pressure. Again we consider systems capable only of pdV work and changes only in internal energy U. For such systems, the First Law is:

$$\delta q = dU + p\, dV$$

For constant p we can write:

$$\delta q = dU + p\, dV = d(U + pV) = dH$$

where $H = U + pV$ is a useful property called enthalpy. We define the *constant-pressure molar heat capacity* C_p as:

$$C_p \equiv (\delta q/dT)_p = (\partial H/\partial T)_p$$

Thus, enthalpy H plays the same role in heating at constant pressure as internal energy U plays in heating at constant volume. Recall again that for 1 mol of an ideal gas, $pV = RT$. Thus:

$$
\begin{aligned}
C_p\, dT = dH &= d(U + pV) = d(U + RT) \\
&= dU + R\, dT = C_V\, dT + R\, dT \\
&= (C_V + R)dT
\end{aligned}
$$

Therefore, $C_p = C_V + R$. For solid and liquid substances $p\, dV$ (at constant pressure) is much less than $R\, dT$ (because dV is so small). Consequently, the difference between C_p and C_V for condensed phases is much less than R.

18. Isothermal Heating.   Now let's consider heating at *constant temperature*. We recognize that processes like melting ice and boiling water are in this category, but we will direct our attention to the heating of a gas obeying $pV = RT$. Once again we rule out changes in potential, kinetic, chemical, electrical, and magnetic energies, so the First Law becomes:

$$
\begin{aligned}
\delta q &= dU + p\, dV \\
&= C_V\, dT + (RT/V)dV \\
\delta q &= 0 + RT\, d \ln V \qquad \text{(for } dT = 0\text{)}
\end{aligned}
$$

where we understand $d \ln V$ as a small or differential change in the natural logarithm of V. Integrating this expression between V_1 and V_2, we find:

$$
Q = RT \ln(V_2/V_1) = W
$$

so that in the *isothermal* ($dT = 0$) expansion of an ideal gas, the heat absorbed by the system Q equals the work done by the system W. In addition, the equation says, as Carnot pointed out, that for a given expansion ratio V_2/V_1, W and, therefore, Q are directly proportional to the temperature T at which the expansion occurs. The higher is the temperature, the greater the amount of work performed.

Isothermal heating's sure
Not to change the temperature.

19. Adiabatic Processes.   Processes occurring in the absence of any heat interaction ($\delta q = 0$) are called *adiabatic*. For our now-common (but constrained) case of $\delta w = p \, dV$ and $dE = dU$, the First Law is:

$$dU = \delta q - p \, dV$$

Let's consider the case of an ideal gas for which $pV = RT$ and $dU = C_V \, dT$. For an adiabatic process $\delta q = 0$ and the equation becomes:

$$C_V \, dT = -p \, dV = \delta w$$

Integration for a finite expansion gives:

$$C_V(T_2 - T_1) = W = -\int_{V_1}^{V_2} p \, dV$$

Thus, we can determine the work done if we know T_2, T_1, and C_V. Sometimes we know V_1 and V_2. To determine the work, we must evaluate the integral $\int_{V_1}^{V_2} p \, dV$. We cannot carry out the integration unless we know how p depends on V as the temperature varies. For $pV = RT$, we can replace p by RT/V in $C_V \, dT = -p \, dV$ and divide both sides by T to obtain for 1 mol of gas

$$C_V \, dT/T = -R \, dV/V$$

Integration of this expression for a finite expansion gives:

$$C_V \ln(T_2/T_1) = -R \ln(V_2/V_1)$$

This relation allows us to determine T_2/T_1 from V_2/V_1. If we know either T_1 or T_2, we can compute the work for an adiabatic expansion from any given V_1 to a given V_2. Because $pV = RT$ throughout the expansion, we can also compute the work if we know p_1 and p_2. We simply substitute $V = RT/p$ in the last expression to obtain (after some algebra):

$$\ln(T_2/T_1) = \frac{\gamma - 1}{\gamma} \ln(p_2/p_1)$$

where $\gamma = C_p/C_V$. (Recall that $C_p - C_V = R$.) Results of such manipulations can be summarized tersely: For adiabatic reversible expansion or compression of an ideal gas, pV^γ, $TV^{\gamma-1}$, and $Tp^{(1-\gamma)/\gamma}$ remain constant.

20. Carnot Cycles and Efficiency.   Finally, we recall the cyclic process invented by Carnot. It comprises in sequence an isothermal expansion, an adiabatic expansion, an isothermal compression, and an adiabatic compression. The extent of each of these processes is adjusted, so that at the end of the sequence, the gas is in exactly the same state as it was in the beginning. In the case of an ideal gas as the working substance, by combining the results of the previous paragraphs we can arrive at:

$$\frac{W}{Q_1} = \frac{T_1 - T_2}{T_1}$$

where Q is the net work done by the gas on the surroundings during the cycle; Q_1 is the amount of heat absorbed from the high-temperature source at T_1 during the isothermal expansion; and T_2 is the temperature of the cold reservoir to which Q_2 is rejected during the isothermal compression. The ratio W/Q_1 is known as the *efficiency* of the cycle. Because each step is carried out reversibly, the cycle can be run backwards. In this case, W will be the work absorbed by the system; Q_1 will be the heat rejected to the high-temperature reservoir; and Q_2 will be the heat absorbed from the low-temperature reservoir.

In this refrigerator mode we are usually concerned with how much Q_2 will be achieved for a given W input. It becomes useful, therefore, to write:

$$Q_2/W = \frac{T_2}{T_1 - T_2}$$

representing the so-called *coefficient of performance* of the Carnot cycle as a refrigerator. Carnot concluded that no cycle could have a higher efficiency as an engine, W/Q_1, than given by the first expression above. If such an engine cycle could be devised, it could be used to drive a Carnot refrigerator. For this super engine, Q_2 exhausted to the low-temperature reservoir would be less than the Q_2 absorbed from that reservoir by the refrigerator. The net result would be equivalent to the spontaneous transfer of heat from a reservoir at low temperature to one at higher temperature. Carnot asserted the impossibility of such a process.

Similar considerations apply to working fluids other than an ideal gas. Thus, stating that heat cannot flow spontaneously from low to high temperature is equivalent to saying that no engine can have a higher efficiency than

Carnot's engine or:

$$W/Q_1 \leqq \frac{T_1 - T_2}{T_1}$$

This statement is *always* true and is one embodiment of the general principle called the *Second Law of Thermodynamics*. It also depends upon the *Zeroth Law* that asserts the existence of temperature and the *First Law* that insists upon the conservation of energy. We will refer to this statement at the *Heat Engine Rule,* or HER.

A boundary helps prevent confounding System content with surroundings!

10 HER Has Much to Say

The 20 statements comprising the last chapter attempted to summarize the basis of thermodynamics as it was understood after a million or so years of human experience with heat and work interactions. No one of those statements was very prepossessing. Nor does the Heat Engine Rule, HER,

$$W/Q_1 \leqq (T_1 - T_2)/T_1 \tag{1}$$

in which they culminated, seem all that imposing. Charlie and Hero had a pretty good working knowledge of heat and work, and they got along without knowing HER. Newcomen and Watt advanced the steam engine art quite effectively long before Carnot probed its fundamental principles. Perhaps the technology that we now "enjoy" could have developed on a purely empirical basis without HER guidance. Whether pure empiricism would be as technologically productive as scientific understanding, or whether science learned more from the steam engine than the steam engine learned from science remain moot questions. But be it a cause or a consequence of technological development, science helps us understand technology's principles and some of their consequences. In this spirit, we will examine HER roles in some of the problems our technological society now faces. We will seek answers to some questions that Charlie and Hero would never even have thought about asking.

HER AT THE POWER PLANT

The public utilities that thrive on society's appetite for electricity have long had a vested interest in what our HER has to say about heat-engine efficiency. Because they sell W in the form of electric power and buy Q_1 in the form of fuel, they are naturally eager to make the ratio W/Q_1 as large as possible. In these days of fuel shortages, their eagerness is even more in earnest. A slightly different form of Equation (1) tells them that according to HER:

$$W/Q_1 \leqq 1 - T_2/T_1 \qquad (2)$$

Therefore, utilities are ever ready to push T_1 to the highest possible levels and to exploit sinks that will provide dumps for their waste heat at the lowest possible T_2 and the lowest possible cost. Some corollaries of HER have in recent years become threats both to the utilities' complacent collection of profits and to our own easy attitudes toward energy "consumption" (rather degradation). It behooves us to take a closer look at HER.

Thermal Pollution   One of the wastes we discharge with abandon is heat. *All* of the energy that comes from falling water and burning fuel is ultimately dissipated in thermal form. In 1979, the environment in the United States had to absorb about 7.8×10^{16} BTU's. That's a lot of heat, but it's still only about $\frac{1}{600}$ of the energy showered on us by the sun. We have a long way to go before our heat output has much direct impact on the *overall* energy balance in the environment. But in some places the *local* heat dissipation rate is already concentrated enough to have a significant effect. The most pressing of the resulting thermal pollution problems emerge at power generating stations. By the present ground rules of the "free" enterprise system, economic factors encourage central stations of ever-increasing size. The net effect, as the power-hungry public insists upon indulging its appetite, is that in some areas waste heat is becoming a problem.

Why "waste" all this heat? The efficiency W/Q_1 that HER has been telling us about is simply the fraction of the total heat released under the boiler that gets "shipped out" as electric power. The rest *must* be rejected to a local

Uxorial the retribution
For this thermal air pollution!

Figure 10-1 At the power plant, Q_2 is waste heat that must be absorbed by the surroundings.

low-temperature sink, usually a lake, river, bay, or other body of water (Figure 10-1). In Equation (2), HER says that the maximum possible efficiency is

$$W/Q_1 = 1 - T_2/T_1$$

Because metals weaken at high temperatures, there is an upper limit at which boiler tubes and other vital components can operate. At present this maximum operating temperature is bout 880 K. For T_2, the temperature at which heat must be rejected, 300 K is a reasonable assumption. If we insert these values in Equation 2, we arrive at an efficiency of about 66 percent. But this is the *maximum possible* efficiency for these temperatures. When all losses are taken into account, the actual overall efficiency of a modern steam plant burning fossil fuel is about 40 percent. Such a plant must discharge locally about 60 percent of all the heat released. It wastes about 50 percent more energy than it produces in the form of electric power.

In the case of nuclear plants, the problem is even more serious. Safety considerations along with structural and heat transfer limitations on the reactor make the effective T_1 for a nuclear plant substantially less than for a plant using "congealed sunshine" as its heat source. Even though the nuclear plant doesn't have as much stack loss (up the chimney), its overall efficiency is so low that it dissipates about twice as much energy in cooling water as it supplies to the power lines in the form of electricity. Moreover, economies of size are more important for nuclear plants. They tend to be much bigger, which means the local discharge of waste heat will be even further intensified as uranium replaces fossil fuel as a heat source.

However, some systems proposed for the future promise higher operating temperatures and, therefore, higher efficiencies. Magnetohydrodynamic generators and direct-fired gas turbines with cooled blades (in which the combustion gases are the working fluid and can be hotter than the cooled walls containing them) are two such proposals. Because of their higher efficiencies, power plants based on these schemes may discharge proportionately less waste heat than conventional generating stations. It will be a long time, however, before any new system can make much of a dent in steam's present domination of the power-generating scene. Any efficiency-spawned relief from present thermal pollution problems is not yet imminent.

Another implication of Equation (2) deserves comment. Relative to the power generated, the amount of waste heat depends directly upon T_2. In summer and in warmer climes when T_2 goes up, the output of waste heat per unit of power output also goes up. And yet it is at just those times and in just those places that air conditioning and refrigeration require more power. Moreover, the tolerance of the local environment for waste heat is likely to be lower. For example, the National Technical Advisory Committee on Water Quality Standards has recommended that in estuaries, the water temperature should be raised by no more than 4°F during fall, winter, and spring, but by only 1.5°F during summer.

There is little doubt that Western society is already at a power-production level that threatens to swamp the cooling capacity of many natural bodies of water. We read in the newspapers of the fish kills that occur when fish are sucked into the cooling-water ducts or up against the screens to keep them out of the pumps. What we read far less about are the probably more devastating effects of raising downstream water temperatures. The concomitant decreases in dissolved oxygen can wreak ecological havoc. These dangers are beginning to be recognized. In some places there are now prohibitions against further discharges of waste heat into local waters. There is also increasing use of cooling towers and ponds in which the waste heat is dissipated by evaporation of water (Figure 10-2). Vaporizing 1 pound of water can absorb 100 times more heat than raising its temperature by 10 degrees on the Fahrenheit scale. But this increased margin, the capital and operating costs of which we will have to pay for in higher power bills, should not make us complacent. By 1990, it is estimated that power plants in the United States alone will transform as much as 100,000 gallons of liquid water into vapor every second! The environmental consequences of such large, local additions to the water in the atmosphere have not yet been sorted out.

Of course, a happier solution to the problem of thermal pollution than expensive cooling towers would be a constructive use for the wasted heat. Space heating of homes, offices, and factories is an option practiced on a limited scale for a long time. Indeed, the city of Västerås in Sweden uses

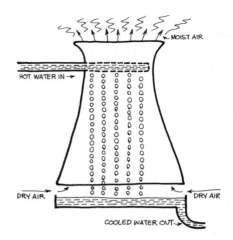

Figure 10-2 Cooling towers take advantage of the high latent heat of water vaporization and are a simple means of dissipating waste heat into the atmosphere.

waste heat to keep streets free of ice and snow in the winter. As fuel costs increase, it will become more and more attractive to invest in the plumbing necessary to distribute hot water or low-pressure steam, especially in densely populated areas and in cold climates. In more temperate zones, unfortunately, there is an unhappy mismatch of supply and demand. On account of air conditioning and refrigeration, the big power load is in the summer when there is the least demand for space heating and, therefore, the smallest outlet for waste heat.

A possibly more intriguing prospect is the use of waste heat to raise the temperature of soil and water to hasten the growth of some useful species of flora and fauna. Still other approaches may emerge as the problem intensifies. But it will require a great deal of time and money to identify, develop, and deploy any new energy system on a scale large enough to solve even partially the waste-heat disposal problem. In the near term, cooling towers will become more familiar features of our landscape.

How Now Brown Out In this age of technologically spawned luxury, the refrigerator and the air conditioner provide comforts to commoners that kings could not once command. These cooling machines are of manifold designs, but they are all inexorably constrained by HER, which in the context of cooling is conveniently expressed:

$$W/Q_2 \leq (T_1 - T_2)/T_2 \tag{3}$$

Here W is the amount of work required to remove Q_2 from a cooled region at T_2 and discharge Q_1 (equal to $W + Q_2$) at the outside temperature T_1 (Figure 10-3).

In the steady state, we become concerned with the *rate* at which heat must be removed from the cooled region to maintain its temperature. If $\dot{Q}_2$ represents that rate (which therefore must also be the rate at which heat leaks back in), then

$$\dot{W} \cong \dot{Q}_2(T_1 - T_2)/T_2 \tag{4}$$

where $\dot{W}$ represents the rate at which work will be required to remove heat at the rate $\dot{Q}_2$, that is, the power consumption of the refrigerator. As we mentioned in Chapter 5, according to Newton's law of cooling, the rate at which heat flows between two objects at different temperatures is proportional to the temperature difference. Thus,

$$\dot{Q}_2 = A(T_1 - T_2) \tag{5}$$

where A is the proportionality constant and $\dot{Q}_2$ is the rate at which heat flows from the outside at T_1 to the inside that is being maintained at T_2 by the refrigerator or air conditioner. We can substitute this expression for $\dot{Q}_2$ into Equation (4) and obtain

$$\dot{W} \cong A(T_1 - T_2)^2/T_2 \tag{6}$$

Let's consider a summer day when the outside temperature T_1 is 86°F (30°C or 303 K). The inside temperature is 77°F, or 298 K. For this condition $\dot{W} = 25A/298$. Now suppose during a heat wave the outside temperature rises to 95°F, or 308 K. To keep the inside temperature at 77°F we find by Equation (6) that $\dot{W} = 100A/298$. In short, a 9°F rise in outside temperature calls for an increase in power consumption by a factor of four! Entirely similar considerations apply to all the refrigerators, freezers, and drinking-water coolers.

Meanwhile, back at the power house, the problem is compounded. Not only are the turbines called upon to generate more power, they find them-

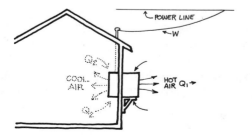

Figure 10-3 In the good old summer time, using a work interaction to cool down is a welcome development. If the indoor temperature is held constant, the heat to be removed Q_2 is equal to the heat that leaks in from outside.

selves handicapped by HER in a direct way. For them

$$\dot{W} \leqq \dot{Q}(1 - T_2/T_1) \tag{7}$$

If we assume that T_2 keeps pace with the heat wave and that the effective T_1 is about 880 K, our 10°F rise in outside temperature (T_1 in Equations (5) and (6); T_2 in (7)) causes a decrease in power-plant efficiency of $\frac{6}{880}$, or about 0.7 percent overall. Actually, the temperature of ponds and streams used for cooling does not change during a heat wave as rapidly as the air temperature. But without exception these heat-sink temperatures go up in summer. Moreover, as the cooling ability of natural sinks becomes saturated, more and more of the cooling load will be born by artificial ponds and towers in which the effective temperature is more closely coupled to the temperature and humidity of the ambient atmosphere. In addition, as nuclear plants take over more and more of the power load the effective T_1 will be lower. Consequently, a given change in T_2 will mean a bigger drop in efficiency.

The decrease in available power is not nearly as dramatic as the increase in demand occasioned by a heat wave, but it certainly comes at a very inopportune time. And we've already mentioned that the restrictions on dumping waste heat are likely to be highest in the summer. All in all, it is no wonder that the brown out is an increasingly frequent warm-weather phenomenon in this air-conditioned society.

Help from the Heat Pump   Thus far, we have been concerned with the difficulties HER causes and the increased cost and inconvenience of those difficulties. Now let's examine a prospective benefit. We will consider the ratio Q_1/W and its role in the relatively unexploited concept, the *heat pump*. Actually, a heat pump performs exactly as a refrigerator. The difference is that we concern ourselves with the heat Q_1 discharged at high temperature T_1 rather than with Q_2, the heat extracted from the region at T_2 that we are interested in cooling. Anyone who has encountered the blast of hot air from an air conditioner's exhaust knows it can be a source of heat as well.

To set the stage, let's assume that we would like to use electric power for space heating in homes and offices. One way to do this is very simple and is shown in Figure 10-4. This arrangement is widely used where electric power is cheap, or where infrequent use justifies the high cost of fuel because of the low cost of equipment. We simply let electric power from the line do work on a resistor. The temperature of the resistor rises. The hot resistor warms the surroundings by heat interaction. Hot plates, toasters, hair dryers, soldering irons, and electric blankets are common examples of the many appliances that operate in this way.

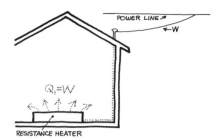

Figure 10-4 Direct electric heat is simple and flexible. The equipment is inexpensive. It is attractive where power is cheap or use is infrequent.

In a steady-state condition, the temperature and the internal energy U of the resistor remain constant. So from the First Law we can write

$$\dot{Q} = \dot{W} \tag{8}$$

where $\dot{W}$ is the rate at which electric power is fed into the resistor, and $\dot{Q}$ is the rate at which the resistor exchanges heat with the surroundings. In short, the amount of heat available from a resistance heater to warm a living room equals the amount of electric power it dissipates. For each kilowatt-hour on our power bill, we obtain about 860 kcal, or 3413 BTU's of heat. That's about the same amount of heat obtainable from burning 3 oz of fuel oil!

Now suppose instead that we use the electric power to "refrigerate" the out-of-doors and discharge the resulting heat into the living room. We simply run a heat engine backward so that heat is absorbed outside and rejected in the house. Figure 10-5 shows such an arrangement. If the heat pump operates at ideal efficiency, we can write

$$Q_1 = W \times T_1/(T_1 - T_2) \tag{9}$$

where T_1 is the house temperature and T_2 is the outside temperature. On a day when it is literally freezing, the outside temperature would be 0°C, or

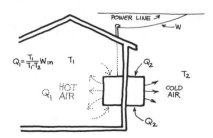

Figure 10-5 The heat pump is simply an air conditioner in reverse. In trying to cool the outdoors, it warms the indoors.

273 K. Inside it would be comfortable at say 294 K, about 70°F. If we insert these values in Equation (9), we find that Q_1 would be $14W$. We obtain 14 times as much heat by using electric power to pump heat in from the outside as we would obtain by dissipating the same amount of power in a resistor!

A gain of 1400 percent in keeping a house warm is not to be sneezed at. Even so, for several reasons, heat pumps are not yet widely used. In the first place, actual operating efficiencies are substantially less than the ideal limit permits. In the second place, a heat-pump system with its associated heat exchangers, compressors, and piping requires a higher initial investment and involves more mechanical maintenance than a resistance heater or the relatively simple furnace required to burn oil or gas. Moreover, even at today's high prices, fuel oil costs only about 1 cent for 1500 BTU's. That same penny in most places would buy only 500 BTU's of equivalent electric power. Thus, one would only break even on fuel costs if the effective gain of the heat pump were three. Purchase and maintenance costs would increase the gain required to break even. Some trials with various combinations of indoor and outdoor temperatures in Equation (9) may persuade you that heat-pump installations would be attractive only in relatively mild climates where electric power is relatively cheap. However, system designs are improving and it turns out that the same machine can run one way as a heat pump in the winter and the other way as an air conditioner in the summer. Consequently, you can expect increasing substitution of this device for furnace–air conditioner combinations, especially where summers are hot and winters are only moderately cold.

Another reason for choosing heat pump installations is emerging. Rising oil and gas prices together with environmental concerns may make them more competitive. Consider two ways of using 15,000 BTU's from burning 1 lb of fossil fuel. If the fuel is burned at a central power station, it would generate about 6700 BTU equivalents of electric power—assuming an overall efficiency in future plants of about 45 percent. Given average inside and outside temperatures on a mild winter day, we have seen that in principle a gain of 14 could be obtained by an ideal heat pump. Let's assume an efficiency of 50 percent, which would cut the gain to 7-fold. We could then expect 46,900 BTU's worth of room heating from 6700 BTU's of electric power. This represents a factor of 3 increase over burning 15,000 BTU's worth of fuel in a home furnace.

As fuel costs go up, this absolute gain is bound to become more and more attractive. Moreover, it is more feasible to provide for clean combustion of low-grade fuels in a big, central power station than in each of many small domestic units scattered over residential areas. Consequently, as tighter and tighter restrictions are imposed on the effluents that combustion systems will be allowed to discharge, the probable trend toward controlled, centralized, combustion systems will enhance the heat pump's prospects.

In these times of rising costs and increasing dependence on uncertain sources of fossil and fissile fuels, heat pumps, waste-heat recovery, improved insulation, and other methods of conservation are beginning to receive the attention they deserve. In the long run, technology may provide us with virtually unlimited quantities of power from nuclear fusion. Even so, we cannot return to profligate normalcy. As populations continue to grow and power consumption continues to climb, we will eventually reach a point where the environment will not absorb waste heat at a greater rate and still remain habitable. When that time comes, we will all have to go on a power diet.

HER AND SCIENCE

In the preceding pages we found that the simple Heat Engine Rule (HER) has some very practical implications. We will now look at some of its more abstract connotations, of interest more to scientists perhaps than to engineers.

The Absolute Temperature Scale   You will recall that during our review of the development of thermometry there was a bothersome dependence of temperature scale in liquid-in-glass thermometers upon the particular liquid used. Alcohol and mercury thermometers agreed at the fixed reference points, as they had to, but not at intermediate temperatures. This inconsistency was alleviated when we went to the gas thermometer. We found that in the limit of very low pressures, all gases provided the same result at all temperatures. But we might still have some doubts as to how fundamental and universal a property temperature is if its value depends in any way upon the nature of the particular thermometric substance used to measure it.

When we considered Carnot's analysis of heat engines, we noted that his conclusions suggested a basis for an absolute temperature scale entirely independent of the particular thermometric substance used to measure it. Let's address this assertion from HER perspectives. Recalling that the First Law requires $W = Q_1 + Q_2$ for any cyclic process in which Q_1 and Q_2 represent all heat interactions and W the net work output, we can rewrite HER as:

$$\frac{Q_1 + Q_2}{Q_1} = \frac{T_1 - T_2}{T_1}$$

Since Q_2 is negative because it represents heat rejected by the engine, we obtain by a little algebraic manipulation:

$$T_2/T_1 = Q_2/Q_1$$

The T's are measured on a gas thermometer scale. The Q's represent heat interactions calculated for a Carnot cycle in which an ideal gas was the engine's working fluid.

But Carnot was able to show that any truly reversible cycle operating between a source at one temperature and a sink at another would have the same efficiency—it would obey HER, no matter what the working substance. Otherwise, we might end up with the spontaneous flow of heat from a cooler body to a warmer one. Thus, if we accept as a law of nature the impossibility of this uphill flow of heat, we must conclude that in a *reversible* heat engine, any working substance (even a mixture of old tires and broken crockery!) would provide the same relationship between Q_1 and Q_2 as shown in the equation when these heat interactions are with reservoirs whose temperatures on the gas thermometer scale are T_1 and T_2. Note that the ratio Q_1/Q_2 would be the same for any reversible engine working with any substance between these two reservoirs even if we measured their temperatures on another scale, for example, the Fahrenheit scale. In that case, however, the ratio of the temperatures would clearly not equal the ratio of the heat interactions.

The important point is that between any two reservoirs at specified temperatures, measured on any scale, Q_1/Q_2 would be the same for all reversible heat engines and all working substances. Consequently, we can conveniently, and arbitrarily, *define* the temperature ratio of these two reservoirs θ_1/θ_2 as being equal to Q_1/Q_2 for a reversible heat engine absorbing heat from one and rejecting it to the other (see Figure 10-6). This procedure is analogous to *defining* the temperature ratio of those reservoirs on the gas thermometer scale as the ratio of the pressures of a fixed volume of gas in thermal equilibrium with the reservoirs (more specifically, the limiting value of the pressure ratio as the gas pressure approaches zero).

Just as in the case of the gas thermometer, asserting that the ratio of any two temperatures is equal to the ratio of values for some observable quantity, in this case Q_1/Q_2, does not completely define a scale. We must also peg the scale to some reference temperature. Again, we choose the triple point of water and give it the same numerical value we did for the gas thermometer: $\theta_{tp} = 273.16$. The resulting temperature scale is identical with the gas thermometer scale we have been using all along; that is, at any temperature, $\theta = T$. The difference is that by resorting to the characteristics

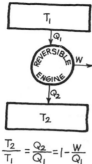

$$\frac{T_2}{T_1} = \frac{Q_2}{Q_1} = 1 - \frac{W}{Q_1}$$

TO FIND T_2/T_1, MEASURE Q_1 AND EITHER W OR Q_2. TO COMPLETE THE DEFINITION OF A SCALE, SPECIFY SOME REFERENCE T, E.G., $T_{tp} = 273.16°$

Figure 10-6 It's important to understand that reversible heat engines can measure temperature, but there's no market for them as thermometers!

of all reversible heat engines using any material as a working medium, we have freed temperature measurement from any dependence upon the characteristic properties of a particular thermometric substance, for example, a liquid of particular composition or a gas in the limit of very low pressure. Because we adopted the symbol T for the gas thermometer temperature in anticipation of its identity with the truly absolute thermodynamic scale just defined, we need not retain the new symbol θ. We will continue to use T for absolute temperatures, perhaps with even more confidence.

In fact, it is unlikely that anyone ever has or ever will actually determine a temperature difference between two objects by measuring the efficiency of a heat engine operating between them. Such a thermometer would certainly seem contrived to Charlie, who had to get by with his sense of touch. Even so, it is reassuring to know that the scale we have been using to measure hotness relates to something fundamental about the nature of an object, independent of the nature of the method by which it is measured. Lord Kelvin was the first to recognize that a reversible heat engine provides the basis for an absolute temperature scale. In his honor, one such scale bears his name.

How About Zero The Kelvin absolute temperature scale and its gas thermometer equivalent provide a zero point at 273.16 degrees below the temperature at which ice, liquid water, and water vapor coexist. It is natural to wonder how matter must behave at a temperature of absolute zero. Many scientists have been very interested in this prospect. They have improved their methods and techniques until now they are able to make observations at temperatures within $\frac{1}{1000}$ of a degree or so of absolute zero. Might we go all

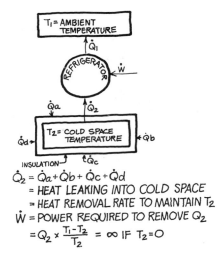

Figure 10-7 How to maintain T at absolute zero. Answer: Harness a source of infinite power. Prospect: No way!

the way? Could we build and operate a refrigerator as shown in Figure 10-7, capable of cooling a body to 0 K and maintaining it at that temperature?

As a practical matter, we would have to envelop the region to be cooled in high-quality insulation. And any heat that leaked through the insulation would have to be continuously removed if the temperature were to be maintained at 0 K. Now HER has something to say. In rate form for refrigerators it becomes

$$\dot{W} \geqq \dot{Q}_2(T_1 - T_2)/T_2 \tag{10}$$

where $\dot{W}$ is the power required to remove continuously the heat flowing at rate $\dot{Q}_2$ through the insulation into the region to be maintained at T_2. T_1 is the temperature of the surroundings that, however small, must be finite. If T_2 is to be zero, then Equation (10) shows that an *infinite* amount of power will be required to overcome any finite heat leak Q_2, no matter how small it is. Unless we can find a perfect insulator, it is impossible to maintain a body at absolute zero even if we could get it there in the first place.

We must also ask how with perfect insulation we would know we had actually reached absolute zero if we could get there. Clearly, there would have to be some sort of thermometer inside the system and some means of communicating its reading to the outside. But any path for flow of information out would also be a path along which heat could travel in, be that path a wire, a tube, or a window for electromagnetic waves. To provide perfect thermal insulation would require blocking all information channels! We would never know when or whether we reached absolute zero unless we had a source of infinite power. In sum, HER makes absolute zero unrealizable and unmaintainable if not unattainable.

How Glaciers Flow   As a final example of HER consequences we examine what happens when water freezes and ice melts. Everybody knows that when water turns to ice, it expands. Otherwise, ice would not float. That this expansion can do work is obvious to anyone who has seen a frozen water pipe burst or a pavement heaved by frost. Let's consider the freezing of 1 g of liquid water having a volume of 1.0001 cm³ at 273.15 K and 1 atm. (Recall that the ice point is slightly below the triple point, 273.16 K.) When the water has turned to ice, the volume will be 1.0907 cm³. This isothermal expansion is shown as line *ab* on the *pV* diagram in Figure 10-8. Because the volume change is so small, only 0.0906 cm³, the scale of the volume coordinate is magnified and the zero volume point has been shifted to the left of the *p* axis. Note that the latent heat of fusion for ice, 79.8 cal/g, has been *rejected* by the water during the freezing process. Recall that while doing work during an isothermal expansion, a gas *absorbs* heat.

After freezing, the water is at point *b* on the diagram. Now we allow the ice to expand adiabatically from *b* to *c* as the pressure decreases. During this step, the volume change is very small indeed. At point *c,* we start warming the ice. As it melts, the volume decreases along path *cd*. The pressure remains constant. The temperature also remains constant during the heat absorption as long as both ice and water are present. Point *d* is the volume just

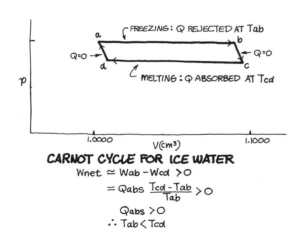

Figure 10-8 One cold-heat engine: A Carnot cycle for ice and water.

as the last bit of ice melts. We now have liquid water at a pressure *lower* than it was at *a*. We complete the cycle by compressing the liquid adiabatically along path *da*. Perhaps by this time you will have realized that we have carried our gram of water through a Carnot cycle comprising an isothermal expansion, an adiabatic expansion, an isothermal compression, and an adiabatic compression in that order. Clearly there are differences between this cycle and the case of the ideal gas analyzed previously. Instead of being absorbed, heat is *rejected* while the system does work on the surroundings. It is *absorbed* while the surroundings do work on the system.

Let's inquire further about conditions during the cycle. In particular, what happens to the temperature? The sequence just described was in a clockwise direction on the *pV* diagram. As we have noted before, this direction means that the enclosed area represents net work done *by* the system. HER requires that if net work is done then heat must have been absorbed at a high tempera-ture and rejected at a lower one. On the diagram, heat was rejected along *ab* and absorbed along *cd*. Thus, HER says that T_{ab} must be lower than T_{cd}. In other words, if we decrease the pressure on ice, we increase its melting temperature; if we raise the pressure, we lower its melting temperature.

We can make this qualitative conclusion a bit more quantitative. Because the volume changes are so small during the adiabatic steps, we can neglect their contribution to the work done during the cycle. Therefore, we can write

$$W_{net} = p_{ab}(V_b - V_a) + p_{cd}(V_d - V_c) \tag{11}$$

Because $(V_b - V_a)$ is very nearly the same as $(V_c - V_d)$, Equation (11) becomes

$$W_{net} = (p_{ab} - p_{cd})(V_d - V_c) \tag{12}$$

The heat absorbed at the higher temperature is Q_{cd}. According to HER,

$$W_{net} = Q_{cd}(T_{cd} - T_{ab})/T_{cd} \tag{13}$$

If we combine Equations (12) and (13), after some rearranging, we obtain:

$$\frac{(T_{ab} - T_{cd})}{(p_{ab} - p_{cd})} = - \frac{(V_d - V_c)T_{cd}}{Q_{cd}} \tag{14}$$

This equation relates the temperature change of the melting point from T_{cd} to T_{ab} and the heat absorbed during melting to the pressure change ($p_{ab} - p_{cd}$).

As we showed in Chapter 7, for a constant-pressure heat interaction between a system and its surroundings, Q can be identified with ΔH, the change in enthalpy of the system. Thus, in more terse and common form, Equation (14) becomes:

$$\frac{\Delta T}{\Delta p} = \frac{dT}{dp} = \frac{T(\Delta V)}{\Delta H} \tag{15}$$

where ΔV is the volume change associated with the enthalpy change ΔH. The minus sign has now disappeared from the right side since we understand ΔH to be the heat of fusion—the heat *absorbed* during freezing (and thus here a negative number because heat is rejected during freezing). If the volume increases during freezing, the sign of the right-hand side is negative (because of the negative ΔH), which means that if Δp is positive, ΔT must be negative. In short, the melting temperature decreases with increasing pressure. If we insert the values associated with water—$\Delta V = 0.0906$ cm³/g and $\Delta H = -79.8$ cal/g $= -(79.8$ cal/g$) (41.4$ cm³ atm/cal$)$ cm³ atm/g—we find that the melting point of ice decreases 0.0075 K/atm. It takes 1000 atmospheres of pressure to decrease the melting point by only 7.5 degrees. This may not seem like a very large effect, but it is enough to let a glacier move.

Recall that pressure is force/area. If the area supporting the force is very small, the pressure can be very large even though the force itself is not. A high-quality record player may exert a tracking force of only 1 g ($\frac{1}{28}$ oz) on the record surface. But if the stylus is only 0.001 inches in diameter at the tip, the local pressure on the record groove will be almost 3000 psi, or about 200 atm. Because of the tremendous mass of ice in a glacier, the pressure at any small projection, like a stone on the underside, is very large and can cause the melting temperature of the ice to go below the glacier temperature. Melting then occurs and the resulting water gets squeezed around the stone. On the "downstream" side of the stone, the pressure is less and the water refreezes. Thus, the ice "flows" over the stone.

You can demonstrate this effect by hanging a wire over an ice cube with a weight at each end as shown in Figure 10-9. The ice melts on the underside of the wire where the pressure is high. The resulting water refreezes as it flows around the wire to the top side where the pressure is low. The wire passes through the ice cube leaving it intact! The heat required to melt the ice on the underside is obtained from the heat released as the water freezes on the top side. This heat is conducted through the wire. Consequently, a wire made of a good thermal conductor like copper will pass through the ice cube faster than a wire made of a poor thermal conductor like nylon.

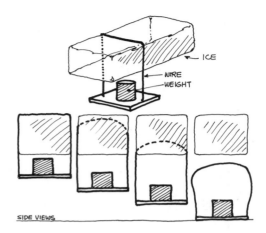

Figure 10-9 How to melt ice by increasing the pressure. The ice becomes water on the underside (high pressure) of the wire, flows around to the top (low pressure) side, and refreezes. Thus the wire passes through the ice but leaves it intact.

SIDE VIEWS

A decrease in freezing point with increasing pressure is unusual in nature. Water is one of the few substances having this property. The metal bismuth is another. It is fortunate that water does behave this way because it means that ice floats. If ice were denser than water, rivers and lakes in all but very warm climes would be all ice all the time except for a thin layer of surface water during the warm seasons! This peculiar property of water helps make ice skating possible. The pressure under the skate blade can be quite high because the weight of the skater is distributed over the relatively small contact area between the blade and the ice. The ice melts and provides a slippery layer of water on which the skater glides. This explanation is clearly not complete. Pressure under the blade on even a very heavy skater would never be high enough to melt ice when the outside temperature is $-20°C$. One supplemental hypothesis is that even at temperatures well below the freezing point, ice is "wet"—has a labile, liquid, surface layer.

Why skating blades so blithely go
E. Clapeyron was first to show.

Vaporization and the Clausius-Clapeyron Equation   Equation (15) also applies to the vaporization of a liquid or solid. In this case, ΔH becomes the enthalpy change due to vaporization (a positive number), ΔV is the difference in volume between a given amount of vapor and the corresponding amount of liquid or solid, and Δp is the change in vapor pressure associated with the change in temperature ΔT. (The vapor pressure of a substance is the pressure of its vapor in equilibrium (coexisting) with a condensed phase at a given temperature T. To make the parallel with the coexisting solid–liquid case clearer, we can think of T as the *boiling* temperature—or, for a solid, the **sublimation temperature**, the temperature at which the vapor pressure is equal to the ambient pressure. When the latter is a standard atmosphere (760 torr), the temperature is called the *normal boiling point*.) When a liquid vaporizes, the volume usually changes a great deal. Consequently, the ratio of temperature change to pressure change is very much larger for coexisting liquid (or solid) and vapor than for coexisting liquid and solid. Moreover, *all* substances expand upon vaporizing. Thus, ΔV is always positive. Therefore, the temperature of a system comprising liquid in equilibrium with vapor *always* increases as the pressure increases. In other words, the vapor pressure of any substance always increases as the temperature increases.

We are usually interested in knowing how the vapor pressure depends upon temperature, rather than vice versa. Therefore, when liquid–vapor or solid–vapor equilibria are concerned, Equation (15) is usually written:

$$\frac{dp}{dT} = \frac{\Delta H}{T\,\Delta V} \tag{16}$$

where we have replaced the ratio of finite differences $\Delta p/\Delta T$ by the ratio of infinitesimal differences dp/dT. In the case of liquid–vapor or solid–vapor systems, the value of ΔV varies substantially with T because of the large change in pressure with T. Thus, for a substantial change in temperature, the right-hand side of Equation (16) will vary considerably over the temperature interval. To calculate the corresponding change in vapor pressure, we must resort to integration.

The relation between pressure, temperature, volume change, and heat of transformation for a phase change was first derived by B. P. E. Clapeyron, a French physicist, in 1834. It may well be the earliest application of thermodynamics to physical chemistry. Consequently, Equation (15) or Equation (16) is widely known as the Clapeyron Equation and, indeed, represents something that science has learned from a study of the steam engine.

Often, the volume change during vaporization is so large that the volume of the condensed phase is negligible relative to its volume in vapor form. Then we can replace ΔV in Equation (16) by V, the volume of the vapor. If we further assume that the vapor behaves as an ideal gas, we can replace V by RT/p and Equation (16) becomes:

$$\frac{dp}{dT} = \frac{p\,\Delta H}{RT^2} \quad \text{or} \quad \frac{dp/p}{dT} = \frac{\Delta H}{RT^2}$$

and then:

$$\frac{d\ln p}{dT} = \frac{\Delta H}{RT^2} \tag{17}$$

The modifications of Equation (16) resulting in Equation (17) were first proposed by Clausius. The latter equation is known as the Clausius–Clapeyron Equation. Multiplying both sides by dT gives

$$d\ln p = \frac{\Delta H}{RT^2}\,dT = -\frac{\Delta H}{R}\,d(1/T) \tag{18}$$

We got the second term on the right-hand side from a legitimate mathematical operation. Recall that the symbol $d(1/T)$ is a very small change in $1/T$ and, therefore, can be regarded as equal to $(1/T_2 - 1/T_1)$ when T_2 and T_1 are very close together. But simple algebra shows that $(1/T_2 - 1/T_1)$ is equal to $(T_1 - T_2)/T_2T_1$. As T_2 and T_1 become very nearly the same, $(T_1 - T_2)$, which equals $-(T_2 - T_1)$, becomes $-dT$, and T_2T_1 becomes T^2. Thus, $d(1/T)$ is equivalent to $-dT/T^2$. Equation (18) can be integrated to give:

$$\ln(p_2/p_1) = -\frac{\Delta H}{R}\left(\frac{1}{T_2} - \frac{1}{T_1}\right) \tag{19}$$

Or, we let the terms involving p_1 and T_1 apply to some reference condition and thus equal a constant, leaving p_2 and T_2 as the only variables. This form allows us to see clearly how vapor pressure depends on temperature:

$$\ln p_2 = \left(\ln p_1 + \frac{\Delta H}{RT_1}\right) - \frac{\Delta H}{RT_2}$$

or, dropping the subscripts and calling the constant A,

$$\ln p = A - \frac{\Delta H}{RT} \tag{20}$$

This expression is the very useful Antoine Equation. The integration assumed that ΔH is constant, usually a fairly good approximation if the temperature change is not too large. In the case of water, for example, ΔH for vaporization of 1 g mol is 9697 cal at 100°C and 10,717 at 0°C, a change of only 10 percent over a temperature interval of 100°C.

The usefulness of the Antoine Equation is due to the linear relationship between $\ln p$ and $1/T$. Knowing the vapor pressure of any substance at two temperatures, one can, with some confidence, construct a straight-line plot of $\ln p$ against $1/T$ and determine the natural log of the vapor pressure at any intermediate temperature. Indeed, one can extrapolate to temperatures modestly above or below the temperatures at which measurements are available. Moreover, the slope of the straight line is $\Delta H/R$, so one can determine the heat of vaporization from vapor pressure measurements at two or more temperatures. Conversely, if one knows the heat of vaporization, a single value of vapor pressure is enough additional information to construct the straight-line plot.

THREE PROBLEMS SOLVED

1. The average January temperature in New Orleans is 56°F, or 516 R. The average July temperature is 83°F, or 543 R. Consider a house in that city that exchanges heat with the outside in accordance with the relation $\dot{Q}_{ex} = A(T_o - T_i)$. (The subscripts o and i stand for outside and inside.) A has the value 600 BTU/hr R. Assume that the inside temperature T_i is to be maintained at 70°F, (530 R) and that electric power costs five cents/kWh. Compute the daily average cost for (a) air conditioning in July if the conditioner's efficiency is 25 percent of the ideal or Carnot efficiency; (b) heating in January with electric resistance heaters; (c) heating in January with a heat pump having the same efficiency as the air conditioner.

Solution

(a) The rate at which heat leaks into the house will be given by $A(T_o - T_i) = 600(543 - 530) = 7800$ BTU/hr or $24 \times 7800 = 187,200$ BTU/day. There are 3413 BTU/kWh, so the heat leak can be expressed as 54.84 kWh/day. For an air conditioner (refrigerator) with a Carnot efficiency of 0.25, $\dot{W} = 4.0 \times \dot{Q}_2(T_1 - T_2)/T_2$. Here T_1 and T_2 correspond to outside and inside temperatures, T_o and T_i, and are expressed on an absolute scale. In this case it is convenient to use the Rankine scale for which $T(R) = t°F + 460$ (actually 459.67). Of course, $\dot{Q}_2$ is the rate heat leaks into the house and must be removed to maintain an inside temperature of 70°F. Thus, $\dot{W} = (4.0)(54.84)(13/530) = 5.38$ kWh/day of electric power required to operate the air conditioner. At 5 cents/kWh, the daily cost would be about 27 cents.

(b) In January, the rate at which heat would leak out of the house is $A(T_i - T_o) = 600(530 - 516) = 8400$ BTU/hr, or 201,600 BTU/day. Divide by 3413 to obtain 59.07 kWh/day. If this heating requirement were met by using a resistance heater, then $\dot{W} = \dot{Q}_1$, and the cost would be $0.05 \times 59.07 = \$2.95$/day. ($\dot{Q}_1$ is the heat lost to outside.)

(c) For a heat pump with an efficiency of 0.25 of the ideal value, we can write: $\dot{W} = (4.0)(\dot{Q}_1)(T_1 - T_2)/T_1 = (4.0)(59.07)(14/530) = 6.24$ kWh/day, or about 31 cents/day. Clearly, in this case the heat pump offers attractive savings in operating costs even though its efficiency is only 25 percent.

2. In Denver, Colorado, which is about a mile above sea level, the barometer reading on a particular day is 600 torr. What would be the boiling point of water on that day?

Solution

(i) Equation (15) can be written for liquid–vapor equilibria as $\Delta T/\Delta p = T(\Delta V)/\Delta H$ where ΔT is the change in boiling point, due to a change Δp in ambient pressure; T is the boiling point on an absolute scale; ΔV is the increase in volume as some amount of liquid vaporizes; and ΔH_v is the heat required to vaporize that amount of liquid.

(ii) In the case of water, T is 100°C (373 K) when $p = 760$ torr (standard atmospheric pressure at sea level). We consider 1 g of water for which ΔH_v is 539 cal at 373 K and ΔV is about 1674 cm³.

(iii) We rearrange the equation slightly to get $\Delta T = T(\Delta p)(\Delta V)/\Delta H_v$. Note that $(\Delta p)(\Delta V)$ has the dimensions of energy as does ΔH_v, so we should express both in the same units. If we express Δp in atmospheres, we get $160/760 = 0.21$ atm. With ΔV in cubic centimeters, the denominator would be in terms of deg × cm³ × atm. Therefore, we should convert ΔH_v into cm³ atm by multiplying by 41.3. Thus, $\Delta T = 373(0.21)(1674)/539(41.3) = 5.61°$. Because the pressure decreased relative to sea level, Δp is negative and so is ΔT. Thus, the calculation gives a boiling point decrease of 5.61 degrees to 94.4°C. The actual value is 93.5°C. Why might our calculation be in error?

3. The vapor pressure of chloroform ($CHCl_3$) at 0°C is 61 torr. At 50°C it is 526 torr. What is the normal boiling temperature? What is the molar heat of vaporization?

Solution

(i) Normal boiling temperature is where the vapor pressure is 1 atm (760 torr). Plugging the given data into Equation (19), we get:

$$\ln(526/61) = \Delta H(1/273 \text{ K} - 1/323 \text{ K})/R$$
$$\ln(8,623) = \Delta H(0.003663 \; \overline{0}.003096)/\text{K } R$$
$$2.154 = \Delta H(0.000567)/\text{K } R$$

so that $\quad\quad \Delta H/R = 2.154/0.000567/\text{K} = 3799 \text{ K}$

$R = 1.98$ cal/g mol K; so $\Delta H = 3799R$ K
$$= 3799 \times 1.987 \text{ cal K/g mol k}$$
$$= 7549 \text{ cal/g mol}$$

(ii) Again we plug into Equation (19) to get:

$$\ln(526/760) = \Delta H(1/T_{NBT} - 1/323\text{K})/R$$
$$\ln(0.692) = \Delta H(1/T_{NBT} - 0.003096/\text{K})/R$$
$$-0.3682 = \Delta H(1/T_{NBT} - 0.003096/\text{K})/R$$
$$-0.3682/\Delta H/R = 1/T_{NBT} - 0.003096/\text{K}$$

From (i) $\Delta H/R$ is 3799 K so that:

$$-0.3682/3799 \text{ K} + 0.003096/\text{K} = 1/T_{NBT} = 0.002999/\text{K}$$

Consequently, $\quad T_{NBT} = 1/0.002999/\text{K} = 333 \text{ K}$

The experimental value for T_{NBT} is 334.7—not bad agreement.

RES MEMORANDA

We have encountered no new information in this chapter. We have simply rearranged old information to obtain new forms. Admittedly, it is not always obvious that the new forms were inherent in what was previously learned. The power of the thermodynamic method rests in its ability to reveal what is inherent but not obvious. Thus, to review what really should be remembered, *reread Chapter 9.* In addition, you might find it interesting to review some of the conclusions reached in this chapter. To wit:

1. Any power-generating plant that we now know how to build and that depends upon heat engines to turn its generators must release locally in the form of waste heat at least as much energy as it sends out in the form of electric power.

2. In general, it should require less energy to transfer heat from a lower temperature to a higher one than to generate from work the same amount of heat at the higher temperature. Therefore, the *heat pump* is a thermodynamically attractive means of space heating.

3. By measuring the heat and work transfers of an ideal heat engine operating between two reservoirs at different temperatures, one can determine the temperature ratio of the reservoirs without regard for the nature of the working substance of the engine. The resultant absolute *thermodynamic temperature scale* is identical with the gas thermometer scale previously defined. The Celsius and Fahrenheit versions of this absolute thermodynamic scale are, respectively, the Kelvin and Rankine scales.

4. It would require an infinite amount of power to maintain a temperature of absolute zero in the absence of an ideal insulator (one that could eliminate all heat interaction between the cooled region and its surroundings). We know of no such insulators. We conclude that absolute zero is unmaintainable and unascertainable if not unattainable.

5. The Clapeyron Equation relates the equilibrium temperature to the equilibrium pressure of any system comprising two coexisting phases, for example, liquid–solid, liquid–gas, or solid–gas. It can be written as $dT/dp = T(\Delta V)/\Delta H$, where dT is the differential change in the temperature of the system associated with a differential change in pressure dp. T is the absolute temperature, ΔV is the change in volume as a quantity of material passes from one phase to the other, and ΔH is the change in enthalpy of the system associated with the phase change of that amount of material. It is numerically equal to the amount of heat interaction required to bring about that phase change. Because water expands upon freezing, this equation says that the melting temperature of ice decreases as pressure increases.

6. In the case of equilibrium between vapor and solid or liquid, it often happens that $V_{vap} \gg V_{sol\ or\ liq}$. If we assume for the vapor that $pV = RT$, we can substitute in the Clapeyron Equation RT/p for ΔV and obtain

$d \ln p/dT = \Delta H/RT^2$—which is known as the Clausius–Clapeyron Equation. If ΔH is constant, usually a good assumption over a reasonable range of temperature, the Clausius–Clapeyron Equation can be integrated to give $\ln p = A - \Delta H/RT$, where A is a constant. This equation is known as the Antoine Equation. It becomes for any pair of temperatures and associated pressures: $\ln(p_2/p_1) = (\Delta H/R)(1/T_1 - 1/T_2)$. In these equations ΔH relates to the transfer of 1 mol of substance from one phase to the other.

MORE OF HER IMPLICATIONS

1. After many years of research, a company perfects a new alloy suitable for making boiler tubes. Its big advantage over previously available materials is that it retains full strength at 1100 K, whereas the maximum working temperature had been 890 K. For a power plant that has cooling water available at 22°C, using the new material for its boilers and turbines would provide what percentage saving in fuel costs?

2. Suppose a power company converts 40 percent of the heating value of the oil it uses into electric power. Suppose also that homeowners can buy heat pumps that are 30 percent as effective as an ideal Carnot machine. If a homeowner maintains the inside of her house at 68°F, at what outside temperature would the heat pump be competitive with an oil furnace that recovered 80 percent of the heating value of the fuel? That is to say, at what outside temperature would the heat delivered by the heat pump per gallon of oil burned at the power company equal the heat delivered by the furnace per gallon of oil?

3. A well-insulated house is heated by an electrically driven heat pump having an overall coefficient of performance based on electrical work input half that of an ideal Carnot machine. When the average outside temperature is −3°C, the monthly electric bill is $100. When the outside temperature averages −23°C, the monthly electric bill is $325. Assume that the electric power cost for all lights and other appliances is $25/month and that the rate of heat loss to the outside (which is what the heat pump must provide) is proportional to the difference between inside and outside temperatures. What is the inside temperature?

4. The temperature in the freezer compartment of a household refrigerator is kept at 0°F. On a summer day the ambient temperature is 92°F. The effective coefficient of performance for the refrigerator (Q_2/W where Q_2 is the heat removed from the frozen food compartment and W is the

electrical work required) is only one-tenth of that achievable by an ideal Carnot refrigerator. If electric power costs 10 cents/kWh (1 watt is 1 joule for 1 second), what is the cost of making one hundred 50-g ice cubes? Assume that the heat of fusion for ice is 80 cal/g and that when the ice cube trays are put in the refrigerator, the water temperature is 77°F. The specific heat capacity of ice is 0.5 cal/g K, half that of liquid water.

5. Home heating oil releases about 144,000 BTU for each gallon burned. Assume that you have a furnace that can release 70 percent of this heat. You have a chance to purchase a heat pump providing 40 percent of the amount that an ideal Carnot machine could produce. Suppose that the average outside temperature during the heating season is 0°C and you want to maintain the inside temperature at 20°C. If electric power costs you 7 cents/kWh, what oil price would make buying the pump attractive? (that is, at what oil price would the operating costs for the two systems be the same?)

6. A room with walls that pass heat at a rate of $B(T_{outside} - T_{inside})$ is cooled by an air conditioner that has half the efficiency of an ideal Carnot machine. It is powered by a 1-kW electric motor that is 90 percent efficient in converting electric power into mechanical work. If B is 700 J/sec K and the outside temperature is 310 K, what is the minimum temperature that can be reached inside the room?

7. Ammonia boils at $-33°C$. Its vapor pressure at 0°C is 4.24 atm. Estimate the molar heat of vaporization and the vapor pressure at 10°C.

8. The vapor pressure of ether at 35°C is 760 torr. Its molecular weight is 74, and its heat of vaporization is 88 cal/g. Find the vapor pressure at 30°C.

9. Glycerin, $C_3H_5(OH)_3$, has a melting point of 20°C at 1 atm pressure. The heat of fusion is 50 cal/g. The density of the liquid is 1.2613 g/cm^3 and that of the solid is 1.2743 g/cm^3. What is the melting temperature at a pressure of 1000 atm?

11   HER Under the Hood

The aggregate work-performing capacity of all prime movers in this country in 1979 was about 3×10^{10} horsepower, enough to stagger Charlie's comprehension. Included are all machines in electric-generating plants, vehicles, homes, factories, mines, boats, airplanes, and farms that use fuel, wind, or water as an energy source. (Electric motors don't qualify. They are run by power generated with prime movers and, thus, are secondary sources of work.) Of this impressive total horsepower only about 2.5 percent was devoted to generating electricity. About 90 percent was under the hoods of rubber-tired vehicles! If we assume that at any one time about 5 percent of those vehicles are on the road and that they are operating at about half their rated output, we find that there is as much power being continuously dissipated on the highways as is being produced in all the electric-generating plants put together. During rush hours and on holiday weekends, the generating plants are left far behind. When we further consider that only about a third of all electricity is for residential use, we must conclude that however hungry for power people may be at home, they are positively ravenous behind the wheels of their cars.

Clearly, it is in cars that most people have their most direct encounters with the conversion of heat into work. It is appropriate, therefore, that we examine the cycles upon which most vehicular power plants are based. Essentially all of them use **internal combustion engines.** High temperatures are attained at an appropriate stage of the cycle, not by means of a heat interaction with an

For transport flexible and fleet
The horseless buggy's hard to beat.

external source, but directly by burning the mixture of fuel and air that is the working fluid of the engine. Heat rejection, or cooling, is simply achieved by expelling the burned gas and replacing it with a new charge of air–fuel mixture. Such an arrangement is attractive because it eliminates the need for a boiler and other heat-transfer equipment. Moreover, because all walls and surfaces in contact with the hot gas can be cooled, for example, the cylinder walls and pistons in a reciprocating engine, the working fluid can be much hotter than the engine structure—not the case when heating and cooling must occur by heat transfer through confining walls, as in a boiler. For these reasons, internal combustion engines are generally both compact and efficient.

THE OTTO CYCLE OR SPARK-IGNITION ENGINE

It has been observed that the French invent, the Germans develop, and the Americans exploit. This sequence certainly seems to apply to the automobile engine. It was a Frenchman, Alphonse Beau de Rochas, who in 1862 proposed the following four-stroke cycle (pictured in Figure 11-1):

1. An *intake* stroke during which the piston descends in the cylinder and draws in combustible mixture through an open inlet valve.
2. A *compression* stroke during which both inlet and exhaust valves are closed as the piston rises and compresses the combustible mixture. Because the compression is fairly rapid it is approximately adiabatic. Therefore, both the temperature and the pressure go up.
3. A *power* stroke during which the hot high-pressure gas—resulting from combustion initiated by a spark at the end of the compression stroke— pushes the piston down and performs work on the crankshaft.
4. An *exhaust* stroke during which the piston pushes the burned gas out through the open exhaust valve.

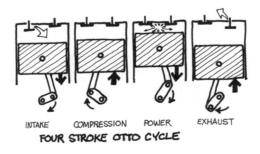

Figure 11-1 The four-stroke Otto cycle—our modern, auto-cycle engine.

INTAKE   COMPRESSION   POWER   EXHAUST

FOUR STROKE OTTO CYCLE

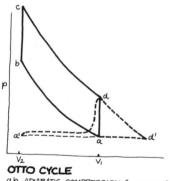

OTTO CYCLE
ab ADIABATIC COMPRESSION (ISENTROPE)
bc CONSTANT VOLUME HEATING (ISOCHORE)
cd ADIABATIC EXPANSION (ISENTROPE)
da CONSTANT VOLUME COOLING (ISOCHORE)
dd'a ACTUAL EXHAUST AND INTAKE, EQUIVALENT
     TO da (NEARLY ISOBARS)
dd'a POSSIBLE ADDITIONAL EXPANSION FOR FURTHER WORK

Figure 11-2 The Otto cycle on a pV (or indicator) diagram. For analysis, the da leg is treated as a constant-volume cooling. In real engines, the equivalent change is achieved by exhausting the burned gas from d to a' and replacing it with fresh fuel–air mixture for a' to a.

Beau de Rochas was a theorist and never built an engine. His ideas were not reduced to practice until 1876 when a German engineer, Nikolaus Otto, built the first prototype. In 1878 manufacture of the engine began in the United States after a patent had been issued to Otto in 1877. Except in retrospect, man has generally admired doers more than thinkers. Possibly for this reason the four-stroke cycle bears the name of its developer, Otto, rather than its inventor, Beau de Rochas. Relatively few of those who drive cars every day have heard of either one.

Let's follow this cycle on the pV diagram shown in Figure 11-2. We start at point a when the piston is at the bottom of its intake stroke and $V = V_1$. As the piston rises during the compression stroke, the state of the gas proceeds along path ab to point b where $V = V_2$. Because the piston moves fairly rapidly there is little time for heat exchange between the gas and the cylinder walls, so the compression is essentially adiabatic. At b a spark ignites the mixture that burns so rapidly that we can neglect the motion of the piston. The combustion thus produces a constant-volume heat addition that causes a rise in pressure and temperature along path bc to point c.

This heat addition is actually the result of a chemical reaction, but in our analysis it is convenient to replace the combustion step by an entirely equivalent heat interaction between the gas and an external source. The piston then begins its downward power stroke during which the gas expands, again nearly adiabatically, along path cd. At point d in a real engine, the exhaust valve opens, the piston rises to expel the burned gas and then draws in fresh mixture so that it is once again at point a. This fairly complicated sequence, which involves an exchange of gases in an engine, is shown by the dashed line $da'a$ in the diagram. However, just as we idealized path bc, we can idealize this complex process of exhaust and intake. It corresponds simply to a rejection of heat at constant volume along path da and could in principle be replaced by heat interaction between the gas and an external sink along solid line da. Thus, the key steps in the Otto cycle are:

1. Adiabatic compression ab.
2. Constant-volume heat addition bc.
3. Adiabatic expansion cd.
4. Constant-volume heat rejection da.

Using the same techniques that we applied in our analysis of the Carnot cycle, let's determine the ideal efficiency of the Otto cycle. Recall that according to Equation (19) in Chapter 8 (see also statement 19 in Chapter 9), the work done by 1 mol of gas during an adiabatic expansion or compression is

$$W = -C_V(T_f - T_i) \tag{1}$$

Therefore, in the Otto cycle shown in Figure 11-2:

$$
\begin{aligned}
W_{ab} &= -C_V(T_b - T_a) \\
W_{bc} &= 0 \qquad \text{(no volume change)} \\
W_{cd} &= -C_V(T_d - T_c) \\
\underline{W_{da}} &= 0 \\
W_{net} &= -C_V(T_b - T_a + T_d - T_c)
\end{aligned}
\tag{2}
$$

We are interested in the efficiency W/Q_{bc}—the ratio of net work to the heat added, or the gasoline burned. For a differential change in a constant-volume process the First Law says

$$dU = C_V \, dT = \delta q \tag{3}$$

Upon integration from initial to final temperature we get

$$Q_{bc} = C_V(T_c - T_b) \tag{4}$$

By combining Equations (2) and (4), we obtain after some rearranging

$$\frac{W}{Q_{bc}} = \frac{T_c - T_b + T_a - T_d}{T_c - T_b} = 1 - \frac{T_a - T_d}{T_b - T_c} \tag{5}$$

We showed in Equation (25) of Chapter 8 that for an adiabatic expansion or compression, TV^{R/C_V} was constant. Therefore, we can write

$$T_b/T_a = (V_1/V_2)^{R/C_V} = T_c/T_d \tag{6}$$

from which we can obtain

$$T_b/T_c = T_a/T_d \tag{7}$$

We subtract 1 from both sides to get

$$T_b/T_c - 1 = T_a/T_d - 1 \tag{8}$$

or

$$\frac{T_b - T_c}{T_c} = \frac{T_a - T_d}{T_d} \tag{9}$$

that gives

$$\frac{T_a - T_d}{T_b - T_c} = \frac{T_d}{T_c} \tag{10}$$

Combining this result with Equations (5) and (6) yields

$$W/Q_{bc} = 1 - T_d/T_c = 1 - (V_2/V_1)^{R/C_V} \tag{11}$$

The ratio of the cylinder volume when the piston is at the bottom of its stroke to the volume when the piston is at the top is known as the **compression ratio** or *cr*. Thus,

$$W/Q_{bc} = 1 - (1/cr)^{R/C_V} \tag{12}$$

This equation says that the efficiency of an Otto cycle engine increases as the compression ratio increases. In the limit as *cr* approaches infinity, the engine efficiency would approach 100 percent! Unfortunately, the cylinder and stroke required to approach an infinite compression ratio would be awkwardly long and the final volume exasperatingly small. Moreover, the resulting final pressure would be hard to contain!

It is interesting to compare the Carnot-cycle efficiency with the Otto-cycle efficiency. For a Carnot cycle operating between the maximum temperature T_c and the minimum temperature T_a, HER tells us that

$$W/Q_c \leqq (T_c - T_a)/T_c = 1 - T_a/T_c \tag{13}$$

The right-hand side of this equation represents the maximum efficiency for any engine operating between these two temperatures. Equation (11) says that for the Otto cycle

$$W/Q_{bc} = 1 - T_d/T_c$$

Thus, the Otto-cycle efficiency is less than the Carnot-cycle efficiency by the amount that T_d/T_c is greater than T_a/T_c. This difference occurs because, in the Carnot engine *all* the heat absorption takes place at the highest temperature and *all* the heat rejection takes place at the lowest temperature. In the Otto engine, heat is absorbed over the temperature interval from T_b to T_c and rejected over the interval from T_d to T_a. In an actual Otto engine, all the heat rejection is in fact at the higher temperature T_d because all the gas at that temperature is discarded through the exhaust valve.

Note that the hot gas at T_d is also at a pressure higher than p_a, the local atmospheric pressure. Consequently, further expansion could perform additional work in an amount represented by the area within the dotted lines drawn between d, d', and a in Figure 11-2.

When airplanes were powered by Otto-cycle engines, this work was frequently recovered by allowing the exhaust gases to expand through a turbine that drove a compressor. The compressor raised the pressure of the gas entering the cylinder during the intake stroke, that is, at a, so there would be more air and fuel burned per cycle. In this way the so-called turbo-supercharger increased the power of the engine over what could be obtained by natural "breathing." This increase in power was particularly important for high altitude flight. Superchargers have occasionally been installed on passenger car engines, but now their use with Otto-cycle power plants is mostly in racing cars. For ordinary use, the cost of the supercharger is not justified, even by the resultant fuel savings from increased efficiency. Don't infer, however, that engine efficiency is not important or that nobody is interested in saving fuel. On the contrary, there has been a continuous effort on the part of automotive engineers to increase engine efficiency. Thus far, they have pursued the more economical approach of increasing the compression ratio as suggested by our analysis. But increasing the compression ratio is not so "simple" after all. We will examine some of its many implications in the following discussion.

THE OCTANE NUMBER GAME

As we saw in Equation (12), the efficiency of an Otto cycle is directly dependent upon the compression ratio. So it would seem desirable to make the compression ratio as high as possible. Mechanically increasing cr is done by reducing the free volume in the cylinder when the piston is at the top of its stroke. Unfortunately, there are some operational consequences of high compression ratio that present problems.

In the ideal cycle we are examining, the heat is added to the working fluid at constant volume. In a real engine the piston is moving and combustion takes time; so heat is added over a small time interval during which the volume is not quite constant. To make most of the combustion (heat addition) occur at the smallest volume, the spark is usually fired slightly before the piston reaches the top of its stroke. Because the flame takes time to propagate throughout the charge, the peak pressure isn't realized until shortly after the piston has started its downward power stroke. We have learned that during an adiabatic compression both the temperature and the pressure increase as the volume decreases. In a gasoline–air mixture, this increase in temperature and pressure causes some preliminary "preflame" oxidation of the fuel.

The products of those preflame reactions burn much more rapidly than the original hydrocarbon molecules. In fact, when the spark ignites this precooked mixture, the flame propagation sometimes is so rapid that a so-called detonation occurs characterized by a shock wave similar to a thunderclap or a sonic boom. The result is an audible ping or knock and an impulsive load on the piston and bearings that can be destructive. Moreover, the peak pressure occurs before the piston reaches top dead center, so there is a loss of power. The onset of knock represents a practical limitation on the compression rate of an Otto engine. There has been a tremendous amount of research aimed at understanding and overcoming the knock phenomenon. It is still not completely understood, but some useful insight has been obtained.

Gasoline is mostly a mixture of hydrocarbon molecules having from 6 to

10 carbon atoms. When the carbon atoms form a straight chain, knock will occur at relatively low compression ratios. When the carbon atoms are arranged in branched chains, much higher compression ratios can be realized before knock becomes a problem. For example, consider normal heptane. If we let H represent a hydrogen atom and C a carbon atom, heptane can be represented by the structural formula

$$
\begin{array}{c}
\quad\ \ \text{H}\ \ \ \text{H}\ \ \ \text{H}\ \ \ \text{H}\ \ \ \text{H}\ \ \ \text{H}\ \ \ \text{H}\\
\quad\ \ |\ \ \ \ |\ \ \ \ |\ \ \ \ |\ \ \ \ |\ \ \ \ |\ \ \ \ |\\
\text{H}-\text{C}-\text{C}-\text{C}-\text{C}-\text{C}-\text{C}-\text{C}-\text{H}\\
\quad\ \ |\ \ \ \ |\ \ \ \ |\ \ \ \ |\ \ \ \ |\ \ \ \ |\ \ \ \ |\\
\quad\ \ \text{H}\ \ \ \text{H}\ \ \ \text{H}\ \ \ \text{H}\ \ \ \text{H}\ \ \ \text{H}\ \ \ \text{H}
\end{array}
$$

Normal heptane is very prone to knock. On the other hand, isooctane

$$
\begin{array}{c}
\quad\quad\quad\ \ \text{H}\quad\quad\ \ \text{H}\\
\quad\quad\quad\ \ |\quad\quad\ \ |\\
\ \text{H}\ \ \text{HCH}\ \ \text{H}\ \ \text{HCH}\ \ \text{H}\\
\ |\ \ \ \ |\ \ \ \ |\ \ \ \ |\ \ \ \ |\\
\text{H}-\text{C}-\text{C}-\text{C}-\text{C}-\text{C}-\text{H}\\
\ |\ \ \ \ |\ \ \ \ |\ \ \ \ |\ \ \ \ |\\
\ \text{H}\ \ \text{HCH}\ \ \text{H}\ \ \ \ \text{H}\ \ \ \text{H}\\
\quad\quad\ \ |\\
\quad\quad\ \ \text{H}
\end{array}
$$

is highly branched and is very resistant to knock. These two compounds represent extreme cases and are the basis for a scale to measure a fuel's knock resistance.

The octane rating scale is constructed in the following way. In a single-cylinder test engine of special design that permits a variable compression ratio (known as a CFR engine, CFR meaning Cooperative Fuel Research), mixtures of normal heptane and isooctane are burned. For each mixture the compression ratio is increased until knock of a specified intensity begins. When this critical compression ratio is plotted against the percent of isooctane in the mixture, a graph like the one shown in Figure 11-3 is obtained. To

Figure 11-3 The octane number scale. The line is determined by tests in an engine with variable compression ratio using mixtures of isooctane with normal heptane. A gasoline that has a critical compression ratio of 6.3, for example, has the same knock resistance as a mixture containing 25 percent heptane and 75 percent octane. Thus, it is assigned an octane number of 75.

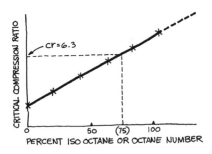

No more power, miles, or speed
With octane higher than you need.

assign a knock rating to any arbitrary fuel blend, one burns a sample in the CFR engine and determines the compression ratio at which knock of the specified intensity occurs.

The **octane number** assigned to the blend is simply the percentage of isooctane in admixture with normal heptane that has equal knock at the same compression ratio. If the blend performs like an 80-20 mixture of isooctane and normal heptane, its octane number is 80. You may have noted that some fuels have octane numbers higher than 100. Such a rating means the fuel resists knock at compression ratios higher than those at which pure isooctane knocks like mad. The actual number is determined by reading off the curve obtained by extrapolating the relation in the graph, that is, the dashed extension of the solid line.

Another important result of research on knock has been the discovery that, when added to gasoline, the compound called "tetraethyl lead" inhibits knock. Dissolved in a gallon of gasoline, a few cubic centimeters of this compound, which has the structural formula (Pb is an atom of lead):

$$
\begin{array}{c}
\text{H} \\
| \\
\text{H}\overset{}{\text{C}}\text{H} \\
| \\
\text{H} \quad \text{H} \ \text{H}\overset{}{\text{C}}\text{H} \ \text{H} \quad \text{H} \\
| \quad\ \ | \quad\ | \quad\ \ | \quad\ | \\
\text{H}-\text{C}-\text{C}-\text{Pb}-\text{C}-\text{C}-\text{H} \\
| \quad\ \ | \quad\ | \quad\ \ | \quad\ | \\
\text{H} \quad \text{H} \ \text{H}\overset{}{\text{C}}\text{H} \ \text{H} \quad \text{H} \\
| \\
\text{H}\overset{}{\text{C}}\text{H} \\
| \\
\text{H}
\end{array}
$$

can markedly increase the octane number. In general it is more expensive to increase the proportion of branched chain molecules in gasoline than to add tetraethyl lead to the straight-run refinery product. Consequently, millions of pounds of tetraethyl lead pass through internal combustion engines every year. The lead emerges as finely divided lead salts that get dispersed in the air we breathe, the water we drink, and the soil in which we grow our food.

The concentration of lead and its compounds in our environment has been steadily increasing since the beginning of the Industrial Revolution. Although we have known about the symptoms and effects of lead, mercury, and other heavy-metal poisoning for a long time, only recently have we learned that continuous exposure to low-concentration levels could be hazardous. Many cases of lead poisoning have been traced to the lead pigments once widely used in paints (and now largely abandoned). There is little evidence that the lead from automotive exhausts poisons people, but the use of tetraethyl lead in gasoline may pose a health hazard. Its use is now decreasing, but before we look at why, let's consider the Diesel engine, the Otto cycle's younger brother that achieves high compression and high efficiency without knock.

THE DOCILE DIESEL

About 25 years after Otto built the first successful spark–ignition engine his compatriot Rudolf Diesel obtained a patent on an engine using the compression–ignition cycle that still bears his name. Diesel's objective had been to develop an engine that could burn powdered coal. Though he failed in that goal, his engine is much less particular about the fuel it uses than the Otto engine. It can burn higher-boiling, lower-grade oils, and it produces more work per unit of fuel consumed. Although many cars are now equipped with Diesel engines, most people know them better as the noisy, stinking, smoking, power plants in trucks, buses, and railroad locomotives. (These engines also have a French connection. Their inventor, though a German, was born in Paris.)

The major operational difference between the Diesel and Otto engines is in how the fire starts. In the Otto engine a mixture of air and gasoline is compressed and then is ignited by an electric spark. In the Diesel engine air alone is compressed adiabatically at perhaps twice the compression ratio used in an Otto engine. The resulting air temperature is so high that when fuel is injected in a spray, it ignites spontaneously. Because the injection, mixing, and burning of the fuel take a relatively long time while the piston is moving down and the gas volume is expanding in the power stroke, the heat addition is effectively at constant pressure instead of at constant volume as in the Otto cycle. The ideal Diesel cycle, therefore, looks a bit different from the Otto cycle and is shown in Figure 11-4. It starts at point a with the piston at the bottom of its stroke. Adiabatic compression occurs from a to b. Fuel injection and burning occur at constant pressure from b to c. The power stroke continues with an adiabatic expansion from c to d.

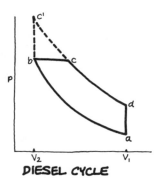

Figure 11-4 The Diesel cycle. The dashed lines show what happens in an Otto cycle with the same compression ratio. The difference occurs because the constant-volume heat addition *bc'* of the Otto engine is replaced in the Diesel engine by the constant-pressure heat addition *bc*.

Then, as in the four-stroke Otto cycle, the piston in a four-stroke Diesel cycle ejects the burned gas in an exhaust stroke and inhales fresh air during an intake stroke to return to starting point a.

In the idealized cycle, this exhaust and intake sequence is equivalent to the constant-volume heat rejection the Otto cycle uses and is represented by the vertical line from d to a. Also, as in the case of the Otto cycle, it is possible to obtain more work by expanding the exhaust gas from d down to atmospheric pressure through a turbine. This recovery step is much more often used with Diesel engines than with Otto engines because efficiency is usually more important in Diesel applications. Diesels are found in trucks, boats, and locomotives where they run more or less continuously. In such cases the cost of the engine relative to the cost of the fuel becomes less important than in an automobile that may run only 5 percent of the time. Consequently, the additional expense of superchargers and other fuel-saving features is more readily justified in Diesel engines. As fuel prices continue to climb, we can expect that automobile engines will also be made more elaborate, and expensive, by fuel-saving features.

The principal difference in principle between the Diesel cycle and the Otto cycle is in how combustion is carried out. In an Otto cycle with the same compression stroke from a to b in Figure 11-4, the combustion–heat addition would occur at constant volume along the dotted line from b to c'. The subsequent adiabatic expansion to d would enclose a larger area on the pV diagram meaning that more work would be performed per cycle. Thus, at the same compression ratio, the cycle efficiency for the Otto engine is greater than the cycle efficiency for the Diesel. So why do Diesels get any attention? Because they can use higher compression ratios and, therefore, achieve higher *operating* efficiencies than Ottos. If an Otto engine were operated at such high compression ratios, the fuel–air mixture would ignite long before the top of the compression stroke. Severe knocking and reduced efficiency would result. Indeed, the engine would probably not run at all. In the Diesel engine fuel is injected only after the air is compressed, that is, at the top of the compression stroke. Thus, there is no chance for precombustion.

It is more appropriate to compare the two cycles on the basis of peak pressure as shown in Figure 11-5. The Otto cycle would then go along path $abcd$. The Diesel cycle path would be $ab'cd$. Clearly, in this case, the area enclosed by the Diesel's path is larger than that enclosed by the Otto's path. It is easy to show that this increase in area means a higher efficiency. From the First Law, we know that $Q_{bb'c} = (U_c - U_b) + W_{bb'c} = Q_{bc(dV=0)} + W_{bb'c}$. In words, the difference between the heat required in the constant-pressure addition of the Diesel cycle and in the constant-volume addition of the Otto cycle equals the difference in the work performed; that is, $W_{ab'cd} - W_{abcd} = W_{bb'c}$. Thus, the efficiency of the Diesel cycle is greater than the efficiency of an Otto cycle at the same peak pressure because:

$$\eta_{\text{Diesel}} = \frac{W_{ab'cd}}{Q_{bb'c}} = \frac{W_{abcd} + W_{bb'c}}{Q_{bc} + W_{bb'c}} > \frac{W_{abcd}}{Q_{bc}} = \eta_{\text{Otto}} \qquad (14)$$

We can write the inequality this way because adding the same amount to both the numerator and denominator of a proper fraction always increases its value. Diesel engines are usually designed for much higher peak pressures than can be tolerated by their Otto cousins. Thus, an even more appropriate comparison would be based on a path for the Diesel of $ab''c'd$, which encloses a much larger area on the pV diagram than a workable Otto could.

Another operating advantage of Diesel engines is that they always start the cycle at the same initial pressure. One adjusts to load variations by controlling the amount of fuel injected into the hot, compressed air. In an Otto engine a throttling valve that varies the inlet pressure is used for such an adjustment. At low loads when the inlet air–fuel mixture is throttled, the average pressure throughout the cycle is low. Under these conditions combustion is usually less complete and the relative heat losses to the walls are much higher. Consequently, the efficiency of an Otto engine decreases rapidly with decreasing load. In a Diesel engine the part–load efficiency remains high.

A third advantage of the Diesel is its willingness to accept lower-grade fuels. For highway vehicles taxes have been, until recently, a major ingredient of fuel price; so the intrinsically lower base-cost of Diesel fuel has been less important than in marine and stationary installments. Because taxes are levied on a per gallon basis, they have not yet gone up in proportion to fuel prices. Thus, base fuel-costs are of much greater relative importance to highway vehicles than they once were.

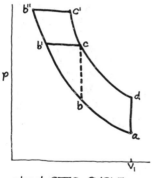

Figure 11-5 Comparison of Otto and Diesel cycles. In practice, Diesel engines operate at much higher peak pressures than Otto engines so that their efficiencies are substantially higher.

abcd OTTO CYCLE
ab'cd DIESEL CYCLE WITH SAME PEAK PRESSURE
ab"c'd DIESEL CYCLE WITH HIGHER PEAK PRESSURE

The ability to use lower-grade fuel is important, but the main attraction of the Diesel cycle is lower fuel consumption. It is not unusual, for example, for a Diesel-powered vehicle to obtain half again as many miles per gallon as its Otto cycle counterpart. There are drawbacks to the Diesel engine. Because it uses high compression ratios and operating pressures, it is inherently heavier and more expensive to build. Therefore, where weight is important, the lighter Otto engine can often win out. Moreover, because of its massiveness, the Diesel does not respond very rapidly. Many motorists would be frustrated at its relatively low acceleration. The differences between these two cousins in the internal combustion engine family are epitomized in two common, but contrasting, transportation applications. The most widely used Diesel engine in railway locomotives has 16 cylinders, weighs 31,000 pounds, and develops 1500 horsepower. One of the best Otto-cycle airplane engines in World War II also developed 1500 horsepower, but it had only 12 cylinders and weighed only 1200 pounds! In spite of their relative mass and sluggishness, Diesel engines are already challenging the dominance of the Otto engine in automobiles. You can bet that many more are on the way.

EFFICIENCY VS. ENVIRONMENT

The inexorable dictate of HER is that high efficiencies in the conversion of fuel to work in any heat engine can only be achieved at high cycle temperatures. Thus, high compression ratios in internal combustion engines make for high efficiencies and more miles per gallon. In the case of the Otto engine, we found earlier that engine knock accompanied high compression ratios. We saw too that these difficulties could be overcome, to some extent, by adding tetraethyl lead to the gasoline or by using hydrocarbons with highly branched carbon chains. The addition of lead to gasoline results in what a growing number of people regard as a pollution hazard. But if we will tolerate somewhat greater expense, the branched chain hydrocarbon content of gasoline can be increased to allow high compression ratios. The Diesel cycle offers another way of realizing high compression ratios and achieving high efficiency. When its plodding performance becomes tolerable, it offers substantial improvements in operating economy.

But the high temperatures inevitably associated with high compression ratios, no matter how they are achieved, have an unpleasant byproduct. They increase the concentration of nitrogen oxides in exhaust gas. In fact, the higher the cycle temperature, the greater is the rate at which nitric oxide is produced. This substance has the formula NO in which N represents a nitrogen atom and O an oxygen atom. After leaving the tailpipe and mixing with ambient air, NO is further oxidized to NO_2, the pungent gas that lends the brownish tinge to smog banks. Ironically enough, nitrogen oxide pollution has become more serious as a result of attempts to decrease another automobile pollutant. Much of the early smog problem in places like Los Angeles had been traced to unburned hydrocarbons in the exhaust gas. Charlie would have blamed his tears on smoke and looked around for the fire. But some careful chemical detective work revealed that these exhaust hydrocarbons under the action of sunlight and atmospheric oxygen, were oxidized to noxious irritants that offended the nose, made the eyes water, and were suspected of much more serious effects. To decrease their hydrocarbon output, engines were tuned for leaner mixtures, so an excess of oxygen would insure more complete combustion of the gasoline. But this excess oxygen then became available for reaction with the nitrogen comprising 80 percent of the atmospheric air that entered the engine. The result has been an increase in the concentration of nitrogen oxides in automobile exhaust as its hydrocarbon content was decreased.

To get out of this bind the Environmental Protection Agency set limits not only on the hydrocarbon emissions but also on the amount of nitrogen oxides that could be released. The automobile manufacturers responded by lowering the compression ratio of their engines and recycling some exhaust gas as a diluent for the incoming air–fuel mixture. The resulting lower peak temperatures reduced the effluent NO_x (shorthand for all nitrogen oxides) but also, as purchasers of new cars in 1974 became painfully aware, substantially reduced miles per gallon. In 1975 catalytic mufflers were introduced to clean up the exhaust by burning the residual hydrocarbons. These mufflers use oxygen in an auxiliary stream of air added to the engine exhaust. They permitted the manufacturers to increase both compression ratio and the fuel–air ratio, and so efficiencies could be restored. However, the muffler catalysts became rapidly poisoned by lead compounds. Lead-free fuel had to be made available. Thus, reduced exposure to lead poisoning in humans becomes a byproduct of the need to avoid lead poisoning in catalytic mufflers!

Most engineers believe that the catalytic muffler is a short-term expedient rather than a long-term solution to the pollution problem. A more attractive alternative would be some form of "stratified charge" engine in which fuel is distributed nonuniformly in such a way as to optimize the combustion process with respect to knock, pollution, and burning efficiency. One model of the Japanese Honda, for example, starts the combustion in an initiation chamber where the mixture is fuel rich and then mixes the resulting burned gas with the rest of the fuel-lean charge in the main cylinder. The exhaust from this engine did not need further treatment (catalytic muffler) until the 1980 vehicle year when more stringent requirements took effect. Further improvements in stratified charge combustion may eliminate "add on" exhaust treatment devices.

Diesel engines can burn hydrocarbons completely if they are not overloaded. But they do produce their fair share of NO_x, and anyone who has followed a heavily loaded truck up a hill knows that irresponsible drivers can make a Diesel engine smoke like a chimney. Recently, it has been claimed that some particulate matter in the exhaust from Diesel engines is carcinogenic. These particulate ejecta are results of the fuel injection and combustion processes peculiar to the Diesel.

Technology may be able to solve these pollution problems and health hazards without abandoning the Otto and Diesel cycles for transportation purposes. Otherwise, we may be forced to develop new means of locomotion, for example, to store electric energy more effectively in portable form and bring back the electric car. Alternatively, another form of heat engine may take over. There are many candidates in the wings. Most of them are

Economy? Without a doubt.
In traffic? Hard to come about!

based on external combustion as in a steam engine. Steady-state, or continuous, burning at atmospheric pressure is much easier to control and does not generally result in such high temperatures. Such combustion promises decreases in both unburned hydrocarbons and nitrogen oxides. On the other hand, in any externally fired system the cycle temperature is limited by the strength of materials to not more than about 900 K, about half the effective temperature in an internal combustion engine. Therefore, efficiencies are bound to be substantially less and engines substantially larger for the same power output. Starting times will be longer, flexibility may be less, and collision hazards may be greater. In spite of all these difficulties, the winds of change are blowing. We may yet live to see a reincarnation of the once-proud steam car! A more likely contender is the Stirling cycle, once common in the hot-air engines widely used for pumping water. This cycle is described in Problems 7 and 8 at the end of this chapter. (The healthiest option would be to take a leaf out of Charlie's notebook and start walking more!)

CLEAN-AIR COST PROBLEM

A relatively recent model of a particular automobile was powered by a 6-cylinder engine with a compression ratio of 9.0. To meet the legal requirements for nitrogen oxides in the exhaust, a later model of the car was fitted with a redesigned cylinder head that provided a compression ratio of 7.0. The earlier model averaged 18 mpg at a highway speed of 50 mph. If gas mileage is directly proportional to engine cycle efficiency, how many miles per gallon could be expected of the new model? Assume C_V for the combustion gases is $2.7R$.

(a) Equation (12) in this chapter gives the efficiency for an ideal Otto cycle as $W/Q_1 = 1 - (1/cr)^{R/C_V}$

(b) If $cr = 9.0$ and $C_V = 2.7R$, then

$$W/Q_1 = 1 - (1/9)^{1/2.7} = 1 - (1/9)^{0.37}$$

To evaluate $(1/9)^{0.37}$ note that it equals $1/(9)^{0.37}$. Recall that $\log x^n = n \log x$. Thus $\log 9^{0.37} = 0.37 \log 9 = 0.37 \times 0.9542 = 0.3530 = 2.25$. Thus $W/Q_1 = 1 - 1/2.25 = 1 - 0.444 = 0.556$.

(c) Similarly, if $cr = 7.0$, then $W/Q_1 = 1 - (1/7)^{0.37} = 0.512$.

(d) We assume that gas mileage is directly proportional to efficiency. Therefore, the new model will get $(0.512/0.556) \times 18 = 16.6$ mpg.

THINGS TO THINK ABOUT WHILE DRIVING

1. In the engines used to power most vehicles heating of the working fluid is achieved by burning fuel–air mixtures inside the cylinder. Cooling consists in exhausting the burned gas and replacing it with a new charge of air–fuel mixture. Such engines are called *internal combustion* engines. They are attractive because they eliminate the need for a boiler and because the working fluid can be much hotter than in systems using heat transfer through some wall.

2. Most automobile engines are based on the Otto cycle, named after the German engineer who first built the 4-stroke engine originally conceived by the Frenchman Beau de Rochas. The Otto cycle comprises an adiabatic compression, a constant-volume heating, an adiabatic expansion, and a constant-volume cooling. The air–fuel mixture is ignited by electric spark. Thus, Otto cycle engines are sometimes called spark–ignition engines.

3. The ratio of the volume of the engine cylinder when the piston is at the bottom of its stroke to the volume when the piston is at the top of its stroke is known as the *compression ratio*. The higher the compression ratio the higher the efficiency. In Otto engines the practical upper limit of compression ratio is determined by the onset of detonative combustion or *knock*. Fuels containing highly branched hydrocarbons and/or a small amount of a compound called tetraethyl lead are more resistant to knock and permit higher compression ratios. *Octane number* is a measure of a fuel's ability to resist knock. The higher this number, the higher is the compression ratio that can be used without encountering knock. It does absolutely no good to use a fuel with a higher octane rating than your car's engine requires.

4. The other common internal combustion engine is based on the cycle developed by the German engineer, Rudolf Diesel. It comprises in order an adiabatic compression, constant-pressure heating, adiabatic expansion, and a constant-volume cooling. In a Diesel engine fuel is injected into the cylinder after the adiabatic compression of the air. The temperature is so high that ignition is spontaneous. Because the fuel burns while it is being injected, there are no accumulations of unburned, air–fuel mixture to detonate and knock. Consequently, Diesel engines can and do operate at much higher compression ratios than Otto-cycle engines. For this reason, and also because their part–load efficiency remains high, Diesel engines are generally more efficient than their

spark–ignition cousins. Moreover, they burn lower-grade, less expensive fuels. Because they are noisy, heavy, and expensive to manufacture, they have not been as widely used in automobiles as in trucks, buses, railroad locomotives, and ships. As fuel prices go up, their use in automobiles seems sure to increase.

5. Practically all present internal combustion engines pollute the atmosphere with unburned or partially oxidized hydrocarbons, carbon monoxide, nitric oxides, sulfur oxides, and lead compounds. Technological advances may solve pollution problems in the near term, but ultimately the engines of future transport may have to abandon the now-common Otto and Diesel cycles.

ENGINE PROBLEMS

1. A particular Otto-cycle engine has a compression ratio of 6 and provides 12 mpg at 40 mph for the vehicle it powers. Assume that the ratio of cycle efficiency to mpg is constant and that the effective C_V for the engine gas is $2R$. How many mpg will be obtained if the compression ratio is increased to 9?

2. Assume that 1 g mol of ideal gas goes through an ideal Diesel cycle in which the compression ratio is 16 and the temperature rise during the constant-pressure heat addition is 1500 K. Assume further that C_V for the gas is $2R$ and that initial T and p are 300 K and 1 atm. Sketch the cycle on a pV diagram. What is the net work output? What is the efficiency?

3. For the same gas properties and initial conditions as in the preceding problem, sketch the diagram, calculate the net work, and determine the efficiency for 1 g mol of ideal gas going through an Otto cycle in which the compression ratio is 8 and the constant-volume heat addition causes a temperature rise of 1500 K.

4. Suppose in the Diesel cycle described in Problem 2 that the constant-volume cooling step is replaced by a continuation of the adiabatic expansion to the initial pressure of 1 atm followed by a constant-pressure compression back to the initial volume. How much additional work would be produced per cycle? (Note that much of this potential increase in work can be realized by letting the exhaust gas pass through a turbine.) How much would the efficiency of the cycle be increased?

5. Suppose the Otto cycle of Problem 3 is fitted with an exhaust turbine that recovers work equivalent to what would be obtained by replacing the constant-volume cooling by adiabatic expansion to 1 atm and constant-pressure compression to initial conditions. How much more work would be produced? What would be the increase in efficiency?

6. A Diesel engine operates on the following cycle, beginning at p_0, V_0, T_0:
   (a) Adiabatic compression to $V_0/25$;
   (b) Constant-pressure heating to $V_0/16$;
   (c) Adiabatic expansion to V_0;
   (d) Constant-volume cooling to T_0. Sketch the cycle on a pV diagram. Compute the change in T for each leg in terms of T_0 and determine the efficiency of the cycle. Assume $C_V = 2R$.

7. The Stirling cycle, named after its inventor, a Scottish minister, comprises in succession an isothermal expansion, a constant-volume cooling, an isothermal compression, and a constant-volume heating. Assume that in a particular engine the isothermal expansion quadruples the volume and the constant-volume cooling halves the pressure. Assume also that C_V is $3R$, that the initial condition of the gas is p_0, V_0, and T_0, and the amount is 1 mol.
   (a) Sketch the cycle on a pV diagram;
   (b) Determine the heat interaction along each leg;
   (c) Determine the overall efficiency of the cycle assuming that any heat absorbed is from a reservoir at the temperature during the isothermal expansion and that all heat rejected is to a reservoir at the temperature of the isothermal compression.

8. If a Carnot engine is operated between reservoirs at the same temperatures used by the Stirling-cycle engine in Problem 7, what would its efficiency be? Explain the difference between the Carnot and Stirling efficiencies. Now assume that the Stirling cycle is operated with a "recuperator" that in effect stores the heat given up by the gas during the constant-volume cooling and uses it to warm the gas during the constant-volume heating step. If the recuperator is ideal it would eliminate all heat absorption and rejection from the reservoirs except during the isothermal expansion and compression. Determine the cycle efficiency for this case and compare it with the Carnot-cycle efficiency. You will then understand why the Stirling cycle with recuperator is being seriously considered for transport applications.

12 Enter Entropy

Our remaining objective in this history of heat will be to arrive at a statement of the Second Law of Thermodynamics that will be somewhat more general (and a lot more abstract) than the HER we have been considering. As a first step in this journey toward generalization we will introduce an enigmatic quantity called *entropy* that has perplexed many generations of students. Just as many generations of teachers have been confounded by attempting to explain what it is. Even so, entropy is such an important and powerful concept that we would be remiss if we did not attempt to understand what in picturesque metaphor has been called "Time's Arrow."

As the discussion unfolds, bear in mind that entropy, like its cousin energy, is neither more nor less than a somewhat contrived quantity that is useful in thermodynamic bookkeeping. Entropy is really no more abstract a concept than energy. It just seems so to most people because it is more unfamiliar. Accordingly, we will spend a fair amount of time defining, describing, and calculating some entropy changes. Charlie would have been bored. Maybe you will be enlightened if not entertained.

PROPERTIES VS. INTERACTIONS

From time to time we have made a point of distinguishing between properties and interactions, the two kinds of quantities with which thermodynamics is preoccupied. In preparation for what is to come, let's review what we mean by these terms.

Properties (quantities such as temperature, pressure, volume, and mass) are simply numbers obtained from carefully prescribed measuring operations. Values of properties are required to describe the condition or *state* of a system. It is important to realize that property values reflect only the present state of a system. They do not relate to its past or its future. Suppose you stick a thermometer in a pot of water on the kitchen stove at 10:00 AM on

*Really just anachronistic
Charlie must think: "hieroglyphic!"*

Monday. Its temperature is 22°C. At the same time on Tuesday morning, the thermometer reads 27°C. All you can say is that the temperature has increased by 5°C. There is no way of telling from the thermometer reading whether during the intervening 24 hours the water had been heated to boiling then cooled, frozen, melted, and then warmed, or had just been slightly warmed. The change in temperature, 5°C, is the same no matter by what path, sequence of states, or succession of values the water temperature rose from an initial value of 22°C to 27°C.

This independence of path is a most important property of properties. It means that for any integration over a particular interval, there is only one possible result. In symbolic form for the case of temperature, this statement becomes:

$$\int_1^2 dT = \Delta T = T_2 - T_1 \tag{1}$$

where $\int_1^2 dT$ means the sum of infinitesimally small changes in temperature dT over the interval from T_1 and T_2. As we have mentioned earlier, dT represents a small change in T such that Equation (1) holds no matter by what path the system goes from T_1 to T_2. It would be more correct to say that T is that quantity for which the integration of differential changes over a given interval will produce the same result no matter how the interval is traversed. Therefore, we use the symbol dT to represent small changes and call it an *exact, total,* or *complete* differential. Quantities like T, which have this character—whose differentials are exact—are what we call properties.

The reason for all this fuss is that there are quantities whose changes in value are not independent of path. Consider the voyage of a ship. We can represent its position at any time in terms of its latitude and longitude. In Figure 12-1, for example, the position at time t_1 is represented by longitude X_1 and latitude Y_1. Two days later at time t_2 the position might be represented by longitude X_2 and latitude Y_2. There are any number of different paths

Figure 12-1 Three possible courses for a ship to sail from X_1Y_1 to X_2Y_2. The change in the ship's position (its "state") is determined solely by the changes in its latitude Y and longitude X (its "properties") and is the same for any course between two particular positions. The distance sailed, on the other hand, depends very much on its particular path.

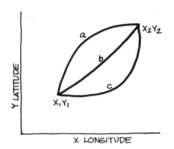

(courses would be the nautical term) by which the ship might have sailed from X_1Y_1 to X_2Y_2. Three possibilities are sketched in the figure. Each has a different length and corresponds to a different distance sailed.

But from just these two sets of coordinates, a navigator would have no way of knowing which of the paths—a, b, or c—had been the actual course. The change in position is exactly the same for each path, but the *distance* traveled is quite different. In short, the position of a ship is its state as characterized by the *properties* of latitude and longitude. Changes in those properties are independent of the path by which the ship goes from one position to another. The distance traveled, on the other hand, is not a property. It depends entirely upon the course between positions.

The path-dependent quantities we have been concerned with are interactions, and the two kinds of interactions thermodynamics recognizes for systems of fixed mass are heat and work. When a system goes from one state to another, the amount of heat absorbed or work done depends very much upon the path or sequence of intermediate states. Consider, for example, the three paths in Figure 12-2 by which a system of 1 mol of ideal gas might go from

Figure 12-2 Three possible paths for the expansion of a gas from p_1V_1 to p_2V_2. The change in all property values and, thus, the change in state, is the same for any path. On the other hand, the work done depends very much on the particular path taken.

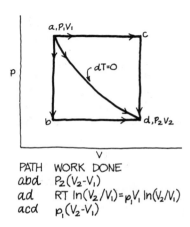

PATH	WORK DONE
abd	$P_2(V_2-V_1)$
ad	$RT \ln(V_2/V_1)=p_1V_1 \ln(V_2/V_1)$
acd	$p_1(V_2-V_1)$

state a at $p_1 V_1$ to state d at $p_2 V_2$. In each case, the work done by the system is given by:

$$W = \delta w = \int_{V_1}^{V_2} p\, dV \qquad (2)$$

This expression is equivalent to the area between the abscissa (V-axis) and the path the system takes from state a to state d. Along the path abd the pressure first decreases at constant volume, and then the volume increases at constant pressure. The area under the curve, and therefore the work, is simply $p_2(V_2 - V_1)$. Along path acd, the work has the much larger value $p_1(V_2 - V_1)$. Path ad is along an isotherm. The work done here is intermediate in value. Remember that it equals $RT \ln (V_2/V_1)$, where RT is itself equal to $p_1 V_1$ or $p_2 V_2$. If you have doubts, review Chapters 4 and 9.

Because the amount of work depends upon path, we use the symbol δw instead of dw for a differential amount of work. Unlike dT, δw is not an exact or total differential. It is equal to $p\, dV$, but this definition is not complete because it cannot be integrated as in Equation (2) without stipulating how p depends upon V. Therefore, δw is an *incomplete*, or *inexact*, differential. Stipulating how p depends upon V is tantamount to defining a path in the pV plane as in Figure 12-2. Thus, $\int p\, dV$ is sometimes called a line, or path, integral because it can be evaluated only along a prescribed path that determines the dependence of p upon V.

Frequently, path-dependent quantities are of great importance. Charlie was well aware that it took more effort to go over a mountain than around it. A shipowner is certainly concerned about the distance his ship sails even though his customer's main interest is in its change of position (from port to port). Likewise, in thermodynamics we are often interested in evaluating interactions along particular paths. Much of the time, however, we obtain some of the information we want about a system change simply from the difference between the initial and final states (changes in property values)— without any reference to a particular path. For example, if we want to know the work actually done when a gas goes from an initial state (p_1, V_1, T_1) to a final state (p_2, V_2, T_2), we must specify the dependence of p upon V during the process, i.e. the path, to evaluate $W = \int_{V_1}^{V_2} p\, dV$. But if we only want to know how much work *could* be done, the First Law tells us that $W = \Delta E + Q = E_2 - E_1 + Q$, that is, the work will equal the energy difference between initial and final states plus whatever heat is absorbed, no matter what the path. The actual work will be path dependent because Q is, but it can be quite useful to know that the work will be determined in part by a property change (ΔE) that is independent of path. It is also noteworthy that the independence of property changes upon path makes them useful in identifying the end points of any particular path, so the integral along that path becomes definite (can be defined and evaluated). Note, for example, the importance of V_1 and V_2 in determining the value of $\int_{V_1}^{V_2} p \, dV$.

For a change of perspective, let's consider your bank account. The deposits and withdrawals are interactions between the account and its surroundings. If you kept track of only the withdrawals, you would have important information on how much money you had spent. But if you ignored the deposits, your information on the state of your account, its balance, would not be very good. In fact, its *balance* is the key property of your account. It is determined by the difference between deposits and withdrawals. Thus, for a small change in the balance B, we can write:

$$dB = \delta D - \delta W \tag{3}$$

where δD represents a small deposit and δW a small withdrawal.

B is a property whose change in value equals the net sum of deposits and withdrawals. Equation (3) relates changes in the *property* balance to the *interactions* of deposit and withdrawal. In fact, both you and the bank determine the balance in your account at any one time by integrating the deposits and withdrawals since the last time the balance was determined. Indeed, it would be extremely difficult for you or your bank to determine your account balance by any direct examination of, or measurement operation on, the bank's contents without reference to these deposits and withdrawals.

It is no accident that Equation (3) is reminiscent of the differential statement of the First Law we have been using:

$$dE = \delta q - \delta w \tag{4}$$

This statement asserts the existence of the property energy by defining its changes in terms of the difference between heat interactions δq and work interactions δw. As we have noted, energy E is an extremely useful quantity because as a system goes from one state to another, its changes are the same no matter what path it travels. Thus, for a substantial excursion, or change:

$$E_2 - E_1 = \int_1^2 dE = \int_1^2 \delta q - \int_1^2 \delta w \tag{5}$$

or

$$\Delta E = Q - W \tag{6}$$

Not only is the property energy *defined* in terms of the dependence of its changes upon heat and work interactions, the changes themselves are nearly always *determined*, in practice, by keeping track of the interactions. Just as in the case of a bank balance, we are usually unable to determine directly the absolute value of a system's energy. Fortunately, it is usually the changes that are important and those can be readily reckoned by Equation (5) from measurable heat and work interactions.

MORE ON PROPERTIES FROM INTERACTIONS

Defining a property and determining its changes by recourse to measurable interactions may confuse you and seem much ado about not very much. But be patient. What we have presented *is* important.

Consider the work interaction between the surroundings and a system of 1 mol of ideal gas. Once again, we can write:

$$\delta w = p \, dV \tag{7}$$

To integrate this equation for a substantial system excursion, we must know how p depends upon V; that is, we must know the path the system follows on the pV plane.

Now suppose we divide both sides of Equation (7) by p to get:

$$\delta w/p = dV \tag{8}$$

Upon integration we obtain:

$$\int_1^2 \delta w/p = \int_1^2 dV = V_2 - V_1 = \Delta V \tag{9}$$

which is independent of the path. No matter how we go from state 1 to state 2, $\int_1^2 \delta w/p$ has the same value, ΔV. By dividing by p, we have changed the inexact differential δw into an exact differential $\delta w/p = dV$. In other words, $\delta w/p$ is the exact differential of a property, volume V. So far no great shakes, but let's continue. Suppose we now divide both sides of Equation (7) by T instead of p. Then, because $pV = RT$, we obtain:

$$\frac{\delta w}{T} = \frac{p}{T} \, dV = \frac{R}{V} \, dV = R \, d \ln V \tag{10}$$

Integration of this equation gives:

$$\int_1^2 \delta w/T = R \ln(V_2/V_1) \tag{11}$$

which is also independent of the path. The value of $\int_1^2 \delta w/T$ is prescribed entirely by the initial and final values of V, not quite so simply as in Equation (9), but just as completely. Dividing a small amount of interaction by T has once again made an incomplete differential δw into a complete, exact differential $\delta w/T$ equal to $R \, d \ln V$.

Neither of these results is very interesting. We already knew about the property defined by $\delta w/p$, and the new property defined by $\delta w/T$ isn't very useful in analyzing processes in thermodynamic systems. But the situation is different if we try these same manipulations with δq. We begin by invoking the First Law:

$$\delta q = dE + \delta w \tag{12}$$

For simple systems in the absence of gravitational, kinetic, magnetic, chemical, electrical, and other effects and for which only $p \, dV$ work is possible,

$$\delta q = dU + p \, dV \tag{13}$$

In the case of an ideal gas, $pV = RT$ so Equation (13) becomes;

$$\delta q = C_V \, dT + \frac{RT}{V} \, dV \tag{14}$$

In integrating this equation over a substantial excursion, we would have no trouble with the first term on the right because C_V is a constant. But to integrate the second term, we would have to know how T depends upon V; we would have to integrate along some particular path in the TV plane. As we very well know, the total amount of heat interaction depends upon the particular path.

Now let's divide both sides of Equation (14) by T:

$$\delta q/T = C_V \, dT/T + R \, dV/V \tag{15}$$

Because R and C_V are both constants, this equation can be readily integrated for any excursion to obtain (see Chapter 4 if you've forgotten):

$$\int_1^2 \delta q/T = C_V \ln (T_2/T_1) + R \ln (V_2/V_1) \tag{16}$$

Clearly, the value of the right-hand side of Equation (16) is determined entirely by the initial and final values of T and V. In other words, the left-hand side $\int_1^2 \delta q/T$ has the same value, no matter what path the system takes from initial state T_1 and V_1 to final state T_2 and V_2. This independence of path means that $\delta q/T$ must be the exact differential of some quantity that is a property of the system.

An important consequence of Equation (16) is what happens to the integral of a property over a cyclic path. Recall that a cyclic process is one that returns a system to its initial state. For such a process, $T_2 = T_1$ and $V_2 = V_1$; so the right-hand side of Equation (16) would vanish. The shorthand statement for this case is $\oint \delta q/T = 0$ in which the circle superposed on the integral sign indicates a cyclic process. This vanishing of the value of an integral around a cycle is adequate evidence that an integrand is an exact or perfect or total differential. As mathematicians like to say, the vanishing of an integral around a cycle is a necessary and sufficient condition for the exactness of its integrand. This kind of evidence for exactness is quite useful because sometimes it is possible to show that the integral around a cycle is zero even when explicit expressions for the result of the integration, as in Equation (16), cannot be readily obtained.

Rudolf Clausius first recognized the nature and importance of the quantity $\delta q/T$. He called the property for which it is the exact differential **entropy** (from the Greek word for transformation) and gave it the symbol S. His christening has endured and we will not argue with a century of custom. Thus, we understand that:

$$\delta q/T = dS \tag{17}$$

which is to be identified with a small change in entropy, a property of a system. It is extremely important to note that each δq is associated with a temperature T for the system as a whole. This association implies that the system has a single, well-defined temperature. Otherwise, we would not know what value of T to use. Furthermore, our assumption that $p = RT/V$ in Equation (14) implies an equally well-defined pressure and volume. Therefore, we conclude that the system is at all times in thermal and mechanical equilibrium and that any change it undergoes comprises a sequence of equilibrium states, each only infinitesimally different from the next. This state of affairs always exists and is implicitly assumed in what we have called *reversible* processes. As we earlier defined them, truly reversible processes are those for which both system *and* surroundings can be restored to their initial conditions. In the argument here we have not been concerned with what may have happened in the surroundings. Consequently, the identification of dS with $\delta q/T$ does not require that the overall process be truly reversible from the perspective of the universe, that is, system plus surroundings. It presumes only that T in the system itself is uniform and can be characterized by a single value.

This distinction between truly reversible processes and those that "seem" reversible to the system is not always clearly drawn. So there is often much confusion about the use of $dS = \delta q/T$. It is valid for any δq experienced by a system that has an identifiable temperature, even though the heating process is irreversible in the universal perspective (e.g., if there is a large difference between T_{system} and $T_{surroundings}$). In sum, Equations (14) through (17) apply directly to any ideal gas system that is in thermal and mechanical equilibrium—that is, has identifiable values for p, V, and T.

The exercise just completed demonstrates the existence of the property entropy for a system comprising an ideal gas. It turns out that $dS = \delta q/T$ for *any* system characterized by a single temperature T whose changes in state are due only to heat interactions. The situation in general is a bit more complicated. We will not try to demonstrate that entropy exists for systems that don't obey $pV = RT$, but you will find such a demonstration in Appendix III. For now we will simply assert that Equation (17) applies to any system at uniform temperature as long as it undergoes no changes other than those produced by a heat interaction. Later we will consider "spontaneous" changes and those due to work interactions. Meanwhile, it may reassure you to know that a century of critical scrutiny and testing has substantiated the applicability of Equation (17) under these limitations, beyond any reasonable doubt. If you can find a bona fide exception, you might well achieve immortality, both literally and figuratively.

You may wonder why we have gone to all this trouble to arrive at the simple generality expressed in Equation (17). The most important reasons will emerge later in this chapter. For now it's enough to show that if we substitute Equation (17) into Equation (13) and do a little juggling, we obtain:

$$dU = T\,dS - p\,dV \tag{18}$$

which is a statement of the First Law for simple systems *entirely* in terms of system properties. The differentials dU, dS, and dV are all exact. Therefore, mathematical manipulation for various processes becomes more straightforward and better defined than if the inexact, and not-very-explicit differentials δq and δw are involved. Note, however, the requirement that the system must at all times be characterized by a uniform temperature and pressure (i.e., it must be at equilibrium) if Equation (18) is to be applicable. Otherwise, it would be impossible to assign values for T and p. Also be reminded once again that the dependence of T on S and p on V must be known if Equation (18) is to be integrated.

COMPUTATION OF SOME ENTROPY CHANGES

The essential message of the last section is that there exists, for any system, a property called *entropy* to which we assign the symbol S. For a small amount of heat interaction q, the associated differential change in entropy dS is equal to $\delta q/T$. We will now use this definition to compute the entropy changes involved in some simple and familiar processes.

The Entropy Change for Melting Ice   Suppose you take a thermos full of ice and water to a picnic on a hot summer day. Because the insulation is imperfect, the ice will slowly melt. And because the ice melts slowly, the temperature in the jug will be fairly uniform at $0°C$. Let's compute the entropy change associated with melting 1 g mol (or 18 g) of the ice. The handbook value for the heat of fusion is 79.67 cal/g, which means about 1434 cal/mol. Thus, we can write:

$$S_{liq} - S_{sol} = \Delta S_{fusion} = \int dS = \int \delta q/T$$

$$= \frac{1}{T} \int \delta q = \frac{Q_{fusion}}{T} = \frac{\Delta H_{fusion}}{T} = \frac{1434}{273} \text{ cal/ g mol deg}$$

$$= 5.27 \text{ cal/ g mol deg} \qquad\qquad (19)$$

As before, $\int dS$ means simply the summation of dS's as they become infinitesimally small and $\int \delta q/T$ means the integration, or summation, of all the $\delta q/T$'s associated with each small quantity of heat interaction δq. The integration is particularly easy in this case because T doesn't change. Therefore, $1/T$ could be removed from under the integral sign to become a simple multiplier of $\int \delta q$, which is the heat of fusion (melting) $Q_f = 1434$ cal/ mol of ice.

*Though it's plainly hard to see
The melting ups the entropy.*

Equation (19) simply means that the entropy of 1 mol of liquid water at 273 K is 5.27 cal/deg greater than the entropy of 1 mol of ice at the same temperature. Conversely, if we extracted enough heat from liquid water at 273 K to form 1 mol of ice at 273 K, the entropy of the system would have *decreased* by 5.27 cal/g mol deg.

Note that all along we have been using absolute Kelvin temperature in the denominator of $\delta q/T$. We could have used the absolute Rankine scale with BTU's as the measure of heat interaction. In those units the entropy change associated with melting 1 lb mol (18 lb) of ice would be 5.27 BTU/deg. Clearly, we cannot use Celsius or Fahrenheit temperatures in the denominator although even advanced students sometimes try it. In this particular case, for example, using the Celsius scale would be absurd because the denominator of $\delta q/T$ would be zero. Note that the units expressing the entropy change are the same as those used for heat capacity—cal/g mol deg or BTU's/lb mol deg. As a matter of simplicity, and to avoid confusion, entropy changes are often expressed in e.u., "entropy units." Thus, the entropy change in melting 1 mol of ice at its normal freezing point is 5.27 e.u. in both the metric and English Engineering systems of units.

The Entropy Change for Vaporizing Water   Another familiar process occurring at a well-defined temperature is the transformation of liquid water to vapor at a pressure of 1 atm. The temperature at which water normally boils is 100°C by definition or about 373 K or 212°F or 640 R. The heat of vaporization at this temperature is 539 cal/g or 9702 cal/g mol. Thus, the entropy change associated with vaporizing 1 g mol of water at normal boiling is:

$$S_{vap} - S_{liq} = \Delta S_{vaporiz} = Q_{vaporiz}/T = \Delta H_{vaporiz}/T$$
$$= \frac{9702 \text{ cal/g mol}}{373 \text{ deg}} = 26.0 \text{ e.u.} \tag{20}$$

This calculation was easy because the temperature does not change during the process.

Note that the entropy change during vaporization of water is about five times as large as the entropy change during fusion. The value of 26 e.u. is a bit on the high side and indicates how unusual a substance water is. Many "normal" (nonpolar) liquids have values of about 21 e.u. for their vaporization entropies. This generalization was first noted empirically by the English physicist, Frederick Trouton, in 1884. Known as Trouton's Rule, it provides a means of estimating the heat of vaporization for a substance *if* you know the temperature at which it normally boils. Just multiply that temperature in degrees Kelvin by the age of majority and you will obtain an approximate value for the heat of vaporization in cal/g mol. If you use the boiling point on the Rankine scale, the result will be in BTU/lb mol.

Entropy Change During Isothermal Expansion of an Ideal Gas   There is another constant-temperature process that we have encountered several times, the reversible isothermal expansion of an ideal gas. In the absence of other than $p \, dV$ work interactions, the First Law for 1 mole of ideal gas can be written (because $dU = C_V \, dT$):

$$C_V \, dT = \delta q - p \, dV \tag{21}$$

and because $pV = RT$, we can write for $dT = 0$ (constant temperature)

$$\delta q = RT \, dV/V \tag{22}$$

We learned how to integrate this expression in Chapter 4, and in this chapter we have already been reminded that the answer is:

$$Q = \int_{V_1}^{V_2} RT \, dV/V = RT \int_1^2 dV/V = RT \ln (V_2/V_1) \tag{23}$$

Because T remains constant, the expression for the associated entropy change is:

$$Q/T = S_2 - S_1 = \Delta S = R \ln (V_2/V_1) \tag{24}$$

The dimensions of R are in cal/mol deg, and the log term is a number, so the units are all right. Thus, there is an entropy increase associated with a volume increase at constant temperature.

Let's reconsider the case of boiling water for a minute. We vaporized 1 g mol. You may recall that the volume of 1 g mol of any ideal gas at 1 atm and 273 K is about 22,400 cm³. At 373 K, this volume would be (373/273) multiplied by 22,400 or about 30,600 cm³. The volume of the liquid before vaporization was about 18 cm^3. Thus, V_2/V_1 is 30,600/18, or about 1700. According to Equation (24), therefore, $R \ln 1700$ is the entropy change associated with the volume change during vaporization of water. Since R has a value close to 1.99 cal/g mol K and ln 1700 is 7.44, the entropy change due to volume increase is about 14.88 e.u. When we computed the total entropy change in the previous section for the overall vaporization process for 1 mol of water, we got 26.0 e.u. We know now that a bit more than half of that entropy change can be associated with the volume change as liquid goes to vapor.

Entropy Changes Due to Temperature Changes   Thus far, all of our computations have been for constant-temperature heat interactions. Now let's consider the more common and slightly more complicated case in which reversible heating brings about a temperature change. If the heating occurs at constant volume, we can write from the definition of constant-volume, specific-heat capacity C_V that $\delta q = C_V \, dT$. Thus:

$$dS = \delta q/T = C_V \, dT/T \qquad (25)$$

Upon integration over a finite change in temperature, we obtain:

$$S_2 - S_1 = \Delta S = C_V \int_1^2 dT/T = C_V \ln (T_2/T_1) \qquad (26)$$

We assumed that C_V doesn't change with temperature; so that it can be removed from in front of the integral sign, it is important to realize that the identification $\int_1^2 \delta q/T$ with $C_V \ln (T_2/T_1)$ releases us from the requirement that the heating process be reversible, that T be uniform and ascertainable throughout the heating process. Now we need to know the system temperature only at the beginning and the end of the heating process. In other words, only the initial and final states must be at thermal equilibrium. What occurs in between is irrelevant.

In the more common and much easier-to-achieve case of heating at constant pressure, we note that $\delta q = C_p \, dT$. By exactly analogous treatment, we then find that:

$$S_2 - S_1 = \Delta S = \int_1^2 dS = \int_1^2 \delta q/T = C_p \int_1^2 dT/T = C_p \ln \frac{T_2}{T_1} \qquad (27)$$

where C_p, of course, is the constant-pressure specific heat capacity. In the case of liquid water, C_p is very close to 1 cal/g deg, or 18 cal/mol deg. If we heat 1 g mol of liquid water from the freezing point at 273 K, to the boiling point at 373 K, the entropy change will be:

$$S_{373} - S_{273} = \Delta S = C_p \ln (T_2/T_1) = 18 \times \ln (373/273)$$
$$= 18 \times 0.315 = 5.66 \text{ e.u.} \qquad (28)$$

Recall that the entropy change for melting 1 g mol of ice was 5.27 e.u. For vaporizing water at 373 K, it was 26.0 e.u. If we consider these three reversible processes to have occurred in sequence to the same mole of water, we can summarize as follows:

1. Ice to liquid water at 1 atm and 273 K: $\Delta S = 5.27$ e.u.
2. Liquid water at 1 atm from 273 K to 373 K: $\Delta S = 5.66$ e.u.
3. Liquid water to vapor at 1 atm and 373 K: $\Delta S = 26.00$ e.u.
    Overall, from ice at 273 K to vapor at 373 K: $\Delta S = 36.93$ e.u.

In other words, as a system of 1 g mol of water goes from ice at 273 K to steam at 373 K and 1 atm, its entropy increases by about 36.9 e.u.

MORE ON THE INDEPENDENCE OF PATH

In each of the immediately preceding computations, we took advantage of the fundamental relation $dS = \delta q/T$ that is true for any system with a uniform temperature undergoing a heat interaction. But what about adiabatic changes? And what about systems that cannot be characterized by a uniform temperature? What happens to the entropy in such cases? In this section we will address ourselves to these important questions.

First let's take another look at the process of melting ice. We found previously that transforming 1 g mol of ice to 1 g mol of water involved an increase in entropy of 5.27 e.u., a number obtained by dividing the molar heat of fusion, 1434 calories, by the melting temperature 273 K. Implicit in this calculation was that we had melted the ice by means of a heat interaction between the container and its surroundings. But suppose we had taken a leaf out of Joule's or Dalton's notebooks and had melted the ice by stirring or rubbing, that is, by a work interaction. If you as an observer had been absent during the melting and had returned to find water where there had been ice, you would have absolutely no way of knowing, and the system would have no way of telling you, whether the melting had been caused by heat or work. The final state, cold water, would be exactly the same in both cases. If we associate an entropy increase with the change of state brought about by a heat interaction, and if entropy is a property whose changes are independent of path, then surely we must associate the same change in entropy with a work interaction that brings about an identical change in state. The same argument applies to the other heating processes we just considered. The identical initial and final states could just as well have been connected by paths along which only work interactions occurred. If the heat interaction caused an increase in entropy, the equivalent work interaction must surely also result in the same entropy increase.

Let's reflect a bit further on this conclusion. As an example, let's say that a system is worked on in the stirring sense of Joule's experiments. The changes that occur are indistinguishable from those brought about by an "equivalent" amount of heat interaction. Joule laid claim to fame (along with the foundations for the First Law) by showing that the amounts of heat and work needed to bring about equal changes always bore the same quantitative relationship to each other. It was thus concluded that because work interactions in mechanics increased the "energy" of a system, the rise in temperature (or the melting of ice or boiling of water) must be associated with an internal energy that can be changed by heat interactions as well as work interactions. Now we have found that the entropy of a system can likewise be *increased* by a work interaction as well as by a heat interaction. What we have *not* found, because it cannot happen, is a *decrease* in the entropy of a system by means of a work interaction. No one has ever observed a case in which the *internal* energy of a system could be decreased solely by a work interaction unless there was also an increase in volume. (More about this later.)

Meanwhile, it is natural to wonder whether a system's entropy is associated with its internal energy. Indeed it is, and we have already seen in arriving at Equation (18) that for simple systems capable only of $p \, dV$ work, the First Law can be written:

$$dU = T \, dS - p \, dV \tag{29}$$

which upon rearrangement gives:

$$dS = dU/T + \frac{p}{T} \, dV \tag{30}$$

For processes at constant volume, the last term vanishes and

$$dS = dU/T = C_V \, dT/T \tag{31}$$

For processes at constant pressure, Equation (30) becomes:

$$dS = d(U + pV)/T = dH/T = C_p\, dT/T \tag{32}$$

You will recall that H, the enthalpy, plays the same role in a constant-pressure process that U, the internal energy, plays in a constant-volume process.

Integration of Equations (31) and (32) for constant-temperature processes gives at constant volume:

$$\Delta S = \Delta U/T \tag{33}$$

and at constant pressure:

$$\Delta S = \Delta H/T \tag{34}$$

If the temperature varies, we obtain equations that we derived earlier. At constant volume:

$$\Delta S = C_V \ln (T_2/T_1) \tag{35}$$

and at constant pressure:

$$\Delta S = C_p \ln (T_2/T_1) \tag{36}$$

All of these expressions for entropy changes can be completely accounted for in terms of changes in other property values between the initial and final states of the system. In general if we know the initial and final values of the other system properties, we can in principle compute the entropy change for any process whatsoever between those sets of values. We can do this even for processes during which $\delta q/T$ may not be definable because the system has no characteristic temperature T or because there has been no heat interaction, so that δq is at all times zero.

Are you somewhat confused? We began by insisting that $dS = \delta q/T$ only for a process in which the system was at uniform temperature. We have concluded that the details of the process make no difference at all and that we can compute the entropy change for *any* process as long as we know the initial and final states of the system—even for a process during which it is impossible to evaluate $\delta q/T$! And indeed that is the case. It is what we mean by "independence of path," and it is the happy and useful property of properties that "they are what they are" without concern for past or future. When we do *not* know a priori what the entropy difference is between two states, we can always determine it by taking the system from the same initial state to the same final state along an actual or imaginary path that permits the computation of $\delta q/T$. Having carried out the computation, we have evaluated the entropy difference between the two states in terms of changes in other properties, and we need never again bother with the $\delta q/T$ unless we happen to forget the answer.

Let's consider a crude analogy. My neighbor has an underground tank holding the fuel oil for his furnace. The tank was installed before he bought the house. Until recently he had no idea how big the tank is or how much oil was in it at any one time. The oil company delivered periodically and he paid the bill. Not long ago he got a long stick that he indexed with lines 1 inch apart. He stopped oil deliveries. Then with his stick he kept watch on the oil level in the tank until there was just enough on the bottom to wet the end of the stick. He called the oil company, and when the truck came out, he persuaded the delivery man to help him "calibrate" his stick. As the oil level rose to each mark on the stick, he recorded how much oil had gone into the tank according to the meter on the truck. Now he can tell how much oil is in his tank at any time by measuring the oil level with his indexed, calibrated stick. No matter what combination of consumption, deliveries, and leaks has led to a particular level (state) (i.e., no matter what the path), he can tell from his stick reading (property of the state) just how much oil he has (another property of the state).

In effect, what my neighbor did with his tank is just what we do in determining entropy. We take the system from one state to another by a process that allows us to add up (integrate) all the $\delta q/T$'s along the way—leading to equations like (32) through (36)—and thus determine the entropy difference between the states. Having done that, by an actual or possible process, we can thereafter determine the entropy difference between those two states simply by measuring other property values—temperature, pressure, volume, mass, and so on—and then applying the equations we have derived.

AN EXTRA EXERCISE

In this chapter we have carried out a number of detailed entropy-change calculations. Even so, let's compute one more entropy change for an actual process, for which we cannot rely on the direct evaluation of $\delta q/T$. We will be

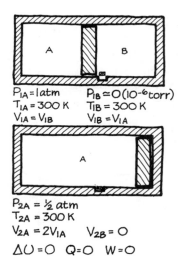

$P_{1A} = 1\,atm$ $P_{1B} \approx 0\,(10^{-6}\,torr)$
$T_{1A} = 300\,K$ $T_{1B} = 300\,K$
$V_{1A} = V_{1B}$ $V_{1B} = V_{1A}$

$P_{2A} = \frac{1}{2}\,atm$
$T_{2A} = 300\,K$
$V_{2A} = 2V_{1A}$ $V_{2B} = 0$
$\Delta U = 0$ $Q = 0$ $W = 0$

Figure 12-3 The free (unresisted) expansion of a gas does no work. If it is also adiabatic, there can be no energy change according to the First Law. In the case of an ideal gas, there is also no temperature change because its energy depends only on temperature, not on pressure or volume.

interested in the result for this particular process anyway, so it won't be time lost for you even if you already understand everything up to this point. Consider a rigid, adiabatic cylinder divided into equal volumes A and B, by a piston locked into place as shown in Figure 12-3. Suppose that on side A, there is 1 mole of gas at atmospheric pressure and a temperature of 300 K. Side B is pumped down to a hard vacuum, so we can ignore the amount of residual gas. (If $P_B = 10^{-6}$ torr, a pressure readily reached by modern vacuum pumps, side B would contain only about 1 billionth of a mole of gas.) The piston is released, and the gas pushes it rapidly into the evacuated region. It bounces off the end wall back into the gas, compressing it to perhaps almost its original value. This process is repeated until finally the piston comes to rest against the end wall. The final pressure is uniform throughout the container at half an atmosphere. The final temperature is the same as the initial temperature, 300 K. (Recall Joule's experiment on free expansion.) What is the entropy change?

Clearly, because the process is adiabatic, $\int \delta q / T = 0$. But just as clearly, T, P, and V are not well defined during the process. Will the entropy change be zero anyway? To answer this question let's imagine that we start all over again with the arrangement shown in Figure 12-4. This time the piston's motion is resisted by a rod connected externally to a cam arrangement so that a weight is raised in the surroundings as the piston moves. For example, the weight could be a mass of water that, initially, just balances the force exerted by the compressed gas in the cylinder. As the water evaporates, the weight diminishes and the cylinder slowly moves to the right. (We made use of this same device back in Chapters 4 and 6.)

228

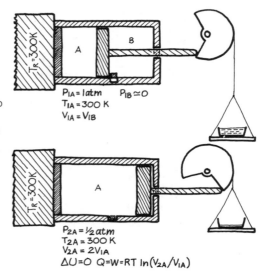

Figure 12-4 An expanding gas will do work if it has to, that is, if its expansion is resisted. In this case, an ideal gas is expanding reversibly; and a constant temperature is maintained by heat interaction with a reservoir at the initial temperature.

We also arrange for the left end of the cylinder to be a thermal conductor in communication with a reservoir at 300 K. Now when we release the piston, it will again come to rest at the end of the cylinder. But in arriving there, it will have performed reversible work on the surroundings. Rather, it will have been the agent by which the gas performed work. As the gas tends to cool by doing work, a heat interaction with the reservoir maintains its temperature at the initial value of 300 K. The net result is that the gas has gone from the same initial state to the same final state as it did during the free, adiabatic expansion in the process we first described. But during the second process, there has been a heat interaction that allows us to calculate ΔS by the integral of $\delta q/T_1$, just as we did in the previous section when we computed the entropy change associated with the heat interaction during the isothermal expansion. Thus, as in the argument that led to Equation (24):

$$\int_1^2 \delta q/T = \Delta S = S_2 - S_1 = R \ln (V_2/V_1) = R \ln 2 \qquad (37)$$

But, in fact, we could have started with Equation (18), which embodies the First Law for simple systems capable only of $p\,dV$ work:

$$dU = T\,dS - p\,dV \qquad (38)$$

Recalling that for an ideal gas $dU = C_V\, dT$ and $pV = RT$, we can readily obtain:

$$dS = C_V\, dT/T + R\, dV/V \tag{39}$$

In the case of the constant-temperature process we have been considering, the first term on the right-hand side disappears. Therefore, integration gives:

$$\Delta S = S_2 - S_1 = R \ln (V_2/V_1) = R \ln 2 \tag{40}$$

which, of course, is the same result as Equation (24) and Equation (37), both obtained by evaluating $\delta q/T$ over the work-producing, heat-absorbing path. In Equation (39), by expressing the entropy change in terms of other properties, T and V in this case, we escape the chore of an accounting job on the particular kind of process from initial to final states that lets us keep track of $\delta q/T$.

There is one more possibility to be considered. Suppose we let the gas in the cylinder in Figure 12-4 expand adiabatically, but that we resist that expansion by forcing the gas to do work slowly and reversibly as in the isothermal expansion just analyzed. In this case, the temperature will decrease while the volume increases, as the First Law requires. But the temperature and volume will have well-defined values at any point in the expansion. Consequently, we can evaluate the entropy change by $\int \delta q/T$ and find that it is indeed zero for this reversible adiabatic expansion. Similar analyses of a wide variety of adiabatic processes lead to the important generality that *reversible* work interactions between a system and its surroundings have no effect upon its entropy. Irreversible work inputs, as in the stirring experiments of Joule, can increase a system's entropy as can heat inputs. But the only way a system's entropy can be *decreased* is by a negative heat interaction, that is, by cooling. There is *no* way that *any* kind of work interaction can *decrease* a system's entropy.

In sum, when a system goes by any process from one state to another, the change in entropy is determined by the changes in more directly observable properties such as pressure, volume, and temperature. When we know the relation between entropy and the other properties, fine. When we don't know it, we can always determine the entropy change by assuming a process path from the same initial state to the same final state along which we can evaluate $\int \delta q/T$. The important qualification is that in the assumed process T must be well defined and uniform throughout the system at every stage.

WORTHWHILE RECOLLECTIONS

The following items include some facts from earlier considerations as well as those introduced in this chapter.

1. Thermodynamic discourse is concerned primarily with relations between and among two kinds of quantities: *properties* and *interactions*.

2. Properties are numbers obtained from prescribed measuring operations. Their values reflect only the state of the system at the time of the measurement. They provide no information about past or future system states. Thus, changes in property values between one system state and another are independent of the path the system takes between the two states. This independence of path means that a small change in a property value, for example, in the pressure, can be represented by the symbol dp—an *exact, total, perfect,* or *complete* differential. These terms mean that $\int_1^2 dp = p_2 - p_1$, a mathematical statement for the independence of path.

3. *Interactions* occur between a system and its surroundings when there is a consistent correspondence between a change in the system and a change in the surroundings. In thermodynamics, there are two kinds of interactions, *heat* and *work*. A small amount of heat is represented by the symbol δq, a small amount of work by the symbol δw. The differential quantities, δq and δw, are *incomplete, inexact,* or *imperfect* differentials. Their integral sums, $\int_1^2 \delta q$ and $\int_1^2 \delta w$, depend upon the particular path by which the system goes from state 1 to state 2.

4. It is possible to relate properties and interactions. For example, in differential form the First Law says that $dE = \delta q - \delta w$. In effect it defines the property *energy,* whose differential dE is exact, as the difference between heat interaction δq and work interaction δW which are inexact differentials.

5. The very important property *entropy* can also be defined in terms of an interaction. In the expression $dS = \delta q/T$, dS is the exact differential of the property S, the symbol assigned to entropy. This statement is true for any system at a uniform temperature that undergoes a change in state due to a heat interaction δq. Reversible work interactions have no effect on entropy.

6. Because entropy is a property, its changes when a system goes from one state to another are independent of the path. When we know the dependence of entropy on the other properties defining the system

state, we need not be concerned with the definition $dS = \delta q/T$. We just compute the entropy change in terms of the changes in other properties. If we don't know how to relate entropy to the other properties, we can always find a path along which we can evaluate $\delta q/T$; and thus determine the entropy change by $\int_1^2 \delta q/T = S_2 - S_1$. Remember that for this relation to apply, the system must at all times be at uniform temperature and that T is to be expressed in terms of either Kelvin or Rankine scales.

7. With the property entropy at our disposal, we can express the First Law entirely in terms of properties. For simple systems capable only of $p\ dV$ work, it becomes: $dU = T\ dS + p\ dV$, where U is the internal energy.

WHAT IN ENTROPY IS THE DIFFERENCE?

1. Compute and compare the entropy increase per gram-mole for fusion and vaporization of each of the following substances.

Name	Formula	Mol. Wt.	t_{melt} (°C)	Q_{melt} (cal/g)	t_{boil} (°C)	Q_{vap} (cal/g)
Hydrogen	H_2	2	-259	13.8	-253	108
Methane	CH_4	16	-183	14.0	-159	138
Chloroform	$CHCl_3$	119	-64	17.6	62	59
n-Heptane	C_7H_{16}	100	-91	33.8	98	76
Ethylene glycol	$C_2H_6O_2$	62	-12	43.3	197	191
Water	H_2O	18	0	79.7	100	540
Benzene	C_6H_6	78	6	30.5	80	94
Naphthalene	$C_{10}H_8$	128	80	35.1	218	76
Sulfur	S_2 (mostly)	64	119	9.2	316	362
Mercury	Hg	200	-39	2.7	357	71
Lead	Pb	207	327	5.9	1620	223
Aluminum	Al	27	659	94.5	1800	1994

2. Exploding a nuclear bomb underground creates a pocket of 10^{10} kg of hot rock at a temperature of 3000 K. Assume that the rock has a heat capacity of 800 cal/kg K and that the earth's crust has the same heat capacity and a temperature of 600 K.
   (a) After the pocket has cooled to the crust temperature, what will have been the entropy change attributable to the cooling process in the pocket?
   (b) in the rest of the earth's crust?
   (c) in the earth as a whole?

3. An iceberg weighing 10^9 kg at a temperature of 0°C drifts into the Gulf Stream, which is at a temperature of 20°C. After several weeks there remains only water at 20°C. What is the entropy change in the universe associated with the disappearance of the iceberg?

4. A cloud at a potential of 10^7 V relative to ground discharges a bolt of lightning that lasts for 0.2 sec and has an average current of 10^5 A. Recall that a current of 1 A through a potential difference of 1 V equals 1 J/s. Assume that all the energy of the lightning bolt is ultimately dissipated as thermal energy, and calculate the amount of entropy produced if $T_{atm} = 300$ K.

5. A meteor at a temperature of 3000 K buries itself in an iceberg that has been floating for some weeks in sea water at 0°C. The meteor weighs 10 kg, has a velocity of 10 km/sec, and a C_p of 800 J/kg K. How much ice will melt? What is the entropy change for the meteor, the iceberg, and the universe?

6. While pushing a piston, 1 kg mol of ideal gas having a C_V of $3R$ doubles its volume at 400 K. The piston is connected with an electric generator. The current from the generator is passed through a resistance heater that heats up 10 kg of water. Assume that one-third of the total work done by the gas is required to push back the atmosphere and the rest heats the water. Determine the temperature change of the water, the entropy change of the water, the entropy change of the gas, and the entropy change of the universe, assuming that all the heat absorbed by the gas comes from a reservoir at 400 K and an initial water temperature of 300 K.

7. One kg mol of an ideal gas goes through a Carnot cycle clockwise on a pV diagram. The temperature of the source is three times the temperature of the sink. Sketch the cycle on a pV diagram and then on a TS diagram. What is the significance of the enclosed area in each case?

8. Recall that the Stirling cycle comprises a pair of isothermal processes connected by a pair of constant-volume processes. (Problem 7, Chapter 11). Sketch a Stirling cycle for 1 mol of ideal gas on pV, TV, TS, and SV diagrams.

9. Consider an Otto-cycle engine in which 1 mol of ideal gas is adiabatically compressed from 1 atm and 300 K to one-eighth of its initial volume, is heated at constant volume through a temperature rise of 1600 K, expands adiabatically to its initial volume, and then cools to its initial temperature. Sketch the process on pV and TS diagrams. Determine the heat interaction and the entropy change for the gas along each leg of the cycle. If the source of heat is a reservoir at 3000 K and the heat rejection during cooling is to a reservoir at 300 K, what is the entropy change of the universe for each cycle of the engine if C_V for the gas is $3R$?

13 Entropy Is the End

Having been introduced to the property entropy, you are now about to explore its role in the most generalized statement of the Second Law yet formulated. Your acquaintance has been so short that you may not feel prepared to test your mastery of what may still seem a mysterious quantity. Be not afraid. If you stumble a bit on the path ahead, you'll be in good company. This last episode in the story that began with Charlie will examine the relation between entropy and the Second Law—in the hope that it may enhance your grasp of this perplexing property that like the cost of living, always and inexorably increases.

IN SEARCH OF ELEGANCE

What we have called the Zeroth Law of Thermodynamics is in effect an assertion that temperature exists. The First Law asserts the existence of energy and requires that it be conserved in all interactions between any system and its surroundings. In contrast to these simple and elegant generalizations, the statements of the Second Law that we have been examining are assertions about what engines and refrigerators can and cannot do. To be sure, we have been able to draw some nonengine conclusions by means of engine arguments, for example, the existence of an absolute temperature scale and the Clapeyron relation about the dependence of freezing and boiling temperatures on pressure. But in the foreground of our considerations, or lurking in the background, there has always been some kind of machinery. Of course, no one can ignore the practical importance of heat engines in providing the particular array of creature comforts that we identify with modern civilization. But this apparent preoccupation with their efficiency lends an aura of crass and clumsy materialism to a generalization that has attained the lofty status of a fundamental principle of nature. The purists among us would surely hope and strive for a statement of the Second Law that would be as terse, general, abstract, and elegant as the First Law statement that says that $dE = \delta q - \delta w$. Therefore, we will now set about cloaking HER in somewhat nobler garb.

Let us note first that our bookkeeping has been done on engines and refrigerators, but our conclusions about the relations between heat and work interactions transcend any particular design or configuration. We started with a particular system (a perfect gas) in a highly specified process (Carnot's cycle). We arrived at relations between heat absorbed and work done that apply to *any* kind of system in *any* kind of cycle. A metal wire whose temperature is alternately raised and lowered would hardly be considered an engine in the ordinary sense. But the net work done during its expansions and contractions bears HER relations to the heat absorbed when it expands. HER governs the relations between heat and work interactions for *any* substance or aggregate of substances in *any* series of changes that ultimately leads back to initial conditions. Because all possible interactions between a system and its surroundings are either heat or work, HER has some very broad implications indeed.

All along we have insisted that HER is true only for a complete cycle or integral number of cycles, that it applies only to cases in which there is no net change in the system comprising the "engine." But for every heat or work interaction the engine undergoes, the surroundings must experience an interaction equal in magnitude but opposite in sign. Every bit of heat *absorbed* by the engine is *rejected* by the surroundings. All the work done *by* the system is *on* the surroundings, and *vice versa*. It follows, therefore, that with appropriate changes in sign, HER applies just as well to the surroundings as to the engine. This observation is particularly important because in this chapter we are going to be as interested in the surroundings—the universe outside the system—as in the system itself.

If a system cycles back to its initial state, any *net* changes resulting from heat or work interactions will have occurred in the surroundings. The engine is merely the agent by which the interactions occur. It retires from the scene when its cycle is complete. Thus, the conclusions reached by analyzing the cycle are related to any changes in the real world that cause, are consequent to, or can be described in terms of, heat and work interactions. There are few, if any, possible changes not falling into this broad class. In short, given all its implications, HER is no mean statement. And we might expect that its message could be embodied in very general form without reference to any particular machine or mechanical device. As you might well suspect, entropy is the key feature of the general form we seek.

ENTROPY CHANGES IN ISOLATED SYSTEMS

The free expansion of an ideal gas in a rigid, adiabatic container is one example of a spontaneous process in an isolated system. By "isolated" we mean no heat or work interactions with the surroundings. As we determined in the last calculation of Chapter 12, the entropy change for an adiabatic, free expansion doing no external work is:

$$\Delta S = nR \ln (V_2/V_1) \tag{1}$$

Because the final volume V_2 is always greater than the initial volume V_1 in any such expansion, the ratio V_2/V_1 is always greater than unity, and the entropy change is, therefore, positive. Incidentally, if in the last calculation in Chapter 12, there had been some gas instead of a vacuum on the right-hand side of the piston, the overall entropy would still have increased as long as the piston was a thermal conductor that allowed final temperature on both sides to be the same. As you can readily determine by a few sample calculations, whenever the initial pressure on the right-hand side is lower than on the left, the increase in entropy of the gas on the left-hand side will be greater than the decrease in entropy of the gas on the right. In short, the equalization of gas pressure in an isolated system characterized by a uniform temperature *always* results in an increase in the entropy of the whole system.

There is another equalization process that will occur spontaneously in an isolated system. Even if you have never actually done the experiment, you would probably be willing to bet that even without stirring, a few grains of salt dropped into a glass of water would ultimately disperse throughout the water—by a process called **diffusion.** Now let's let this diffusion process occur inside a rigid, adiabatic cylinder containing a piston that is permeable to water but doesn't pass dissolved salt. (Such semipermeable membranes are available and, in fact, are used in what is known as the *reverse osmosis* process for desalting sea water.) As indicated in Figure 13-1 the undissolved salt is initially placed in the small volume between the piston and the left end

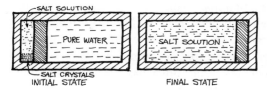

INITIAL STATE   FINAL STATE

Figure 13-1 Because the piston is permeable to water and not to salt, the salt molecules tend to migrate into the pure water and the pure water to migrate into the salt solution—resulting in an *osmotic pressure* difference across the piston that moves it in the direction of pure water.

236

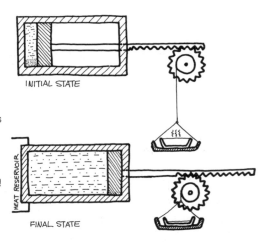

Figure 13-2 Osmotic pressure can be harnessed to do reversible work. Heat interaction with a reservoir ensures constant temperature; so the final state is identical with that in Figure 13-1. This diffusion process is very slow; but, in principle, with the oceans as a source of salt water and the world's rivers for fresh water, osmotic-pressure work could provide the world's power needs!

INITIAL STATE

FINAL STATE

HEAT RESERVOIR

of the cylinder. It dissolves and diffuses through the water but cannot pass through the piston. But the diffusion process continues because more water molecules pass through the piston from right to left than vice versa. The piston thus moves to the right, impelled by what we call *osmotic pressure,* the often considerable pressure we would have to exert on the right-hand side of the piston to keep it from moving. For example, the concentration of salts in sea water is only about 3.5 percent by weight. When the salt concentration in the water on the left-hand side of the piston is at this level, the resulting osmotic pressure would be over 22 atmospheres! If the piston is unrestricted in its motion, it will simply move slowly to the end of the cylinder and come to rest as shown in the sketch. Because the cylinder is rigid and adiabatic, there can have been no heat or work interactions with the surroundings. Therefore, the First Law requires that the total energy of the system be unchanged.

Now we repeat the experiment except that we resist the motion of the piston with a rod and cam arrangement that permits reversible work to be done on the surroundings by the motion of the piston. Figure 13-2 shows a possible arrangement using evaporating water, much like the one we illustrated in Figure 12-4. In this case, the total energy of the system will decrease by an amount equal to the work done. However, we can offset this decrease by means of a heat interaction that ensures a constant temperature. Thus, the final state of the system is the same as in the first experiment when the piston did no work. Clearly, by the latter process the system's entropy is increased by the heat interaction. Therefore, because the initial and final states are the same for both processes, there must have been an equal entropy increase associated with the first process, which we identify with the unrestricted diffusion of salt to produce a uniform concentration throughout the system. Enough experiments of this kind or their equivalent have been done to justify the conclusion that in an isolated system, any equalization of concentration differences of dissolved species is accompanied by an increase in system entropy.

There is one more important equalization process to examine. Consider two identical blocks of copper, a and b, each with heat capacity C. We put both blocks in a rigid, adiabatic enclosure at a time when T_a is 400 K and T_b is 200 K. The arrangement is shown in Figure 13-3. To analyze this situation, we will treat each block as a subsystem and assume that the system as a whole is the sum of the subsystems. After sufficient time, the blocks will be at the same temperature. Because the heat capacities are the same, it is clear that the final temperature will be the arithmetic mean of the initial temperatures (or 300 K). Because the volume change of copper with temperature is small, we can assume that the heating process is at constant volume.

Therefore, from Equation (26) in Chapter 12 (obtained by integrating $dS = \delta q/T = C_v dT/T$):

$$\Delta S_a = C_V \ln (T_{2a}/T_{1a}) = C_V \ln (300/400)$$
$$\Delta S_b = C_V \ln (T_{2b}/T_{1b}) = C_V \ln (300/200)$$

$$\Delta S_{tot} = \Delta S_a + \Delta S_b = C_V \left(\ln \frac{T_{2a}T_{2b}}{T_{1a}T_{1b}} \right) = C_V(\ln \tfrac{3}{4} + \ln \tfrac{3}{2})$$

$$\Delta S_{tot} = C_V \ln \tfrac{9}{8} \qquad (2)$$

Because $\tfrac{9}{8}$ is greater than unity and C_V is positive, the total entropy of the system was increased by the temperature equalization. Such a heat interaction between two objects at different temperatures *always* results in an entropy increase for the system as a whole. The differential statement for any small heat interaction during the temperature equalization process is:

$$dS_{tot} = dS_a + dS_b = (\delta q/T)_a + (\delta q/T)_b \qquad (3)$$

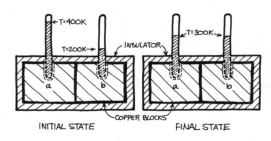

INITIAL STATE   FINAL STATE

Figure 13-3   Spontaneous thermal equilibration in an isolated system comprising two copper blocks. It *always* happens.

If T_a is greater than T_b, δq_a will be negative and δq_b will be the same size but positive. The negative term has the larger denominator T_a, so the positive term is larger. Thus, dS_{tot} is positive. If T_b is greater than T_a, *mutatis mutandis,* and you get the same result. In short, the flow of heat from a hot object to a cooler one always results in an increase in the total entropy of the two.

We have now considered processes in isolated systems by which initial differences in temperature, pressure, and concentration have been equalized. On the basis of our observations, and because there have been centuries of corroboration and not a single exception, we risk the bold generalization that in any isolated system, the equalization of differences in temperature, pressure, or concentration will always result in an increase in entropy of the system. *Never* has it been observed that any of these processes can go the other way. Never has an initially uniform, isolated system ever spontaneously developed nonuniformities in temperature, pressure, or concentration.

In addition to these spontaneous equalization processes, there is another class of phenomena that can produce changes in isolated systems. These might be termed "energy transformations." Here are some examples:

1. An electric motor with a flywheel on its shaft is enclosed in a rigid, adiabatic housing. The motor is energized from an external power supply until a high rotational speed is reached. The electrical circuit is then broken, and the system is completely isolated. After a period of time, whose length depends upon how frictionless the bearings are, the motor and flywheel come to rest. Careful observation shows that the temperature of the entire system is higher than when it was first isolated.

2. A ball high inside an evacuated, rigid, adiabatic enclosure is released and starts bouncing. After a while, it comes to rest on a plate at the bottom of the enclosure. Careful measurements show that the temperature of the plate and the ball have increased slightly.

3. A gaseous mixture of hydrogen and oxygen is contained within a rigid, adiabatic bomb that also contains several liters of liquid water. Also in the bomb is a clockwork mechanism that at a preset time breaks a small glass tube containing a piece of platinum. Catalyzed by the platinum, the hydrogen and oxygen react. After the reaction is over, inspection of the bomb reveals that there is slightly more water, no hydrogen and oxygen, and a somewhat higher temperature than at the beginning.

4. An electric capacitor connected to a resistance wire is enclosed in a rigid, adiabatic envelope. After the envelope is sealed, a clockwork mechanism closes the circuit and the capacitor discharges through the resistance.

After the discharging is over, the temperature of the contents of the envelope is slightly higher than it was to begin with.

5. A highly extended spring has one end attached to a rigid support and the other to a block resting on a rough surface. Initially, the block is secured by a string fastened to another rigid support. The string breaks, and the block slides with friction until the spring is no longer under tension. Inspection shows that the temperature of the spring, block, plane, and supports has increased.

6. A container of radioactive waste is sealed into a concrete block that is then buried deep in a glacier whose surface temperature remains constant at 0°C. Several years later, investigation reveals that deep within the glacier the radioactive block is surrounded by a pool of liquid water at 0°C.

In each of these scenarios, there has been an observable, spontaneous change within an isolated system. Moreover, precisely the same changes could also have been brought about (in principle) by a combination of reversible work and heat interactions with the surroundings. For instance, the electric motor and flywheel could have been brought to rest by doing shaft work on the surroundings, for example, by raising a weight. Then a heat interaction with the surroundings could have been used to change the temperature that in the isolated case was changed by frictional dissipation in the bearings. Similarly in the third example, the hydrogen and oxygen could have been reacted in a fuel cell of the type used to generate electric power in spacecraft. The resulting power could have done work on the surroundings. Meanwhile, a heat interaction with the surroundings could have raised the temperature inside the bomb. Even in the case of the radioactive waste, it would have been possible, even if awkward, to convert some of the energy released by nuclear reaction into external work.

Thus, each of these systems could have been brought to the same final state by a reversible path involving absorption of heat from the surroundings. Along that path the system would have gained entropy. Any entropy change is independent of the system's path between initial and final state. Therefore, each of these spontaneous processes within an isolated system caused an increase in the entropy of that system.

Every one of the six processes just examined involved the generation within the system of *internal* energy at the expense of energy in some other form. By a more formal and more terse argument, we can arrive at the same conclusion. First, we note that a differential change in the total energy of a system can be considered as the sum of differential changes in all the possible components of the total energy. Thus:

$$dE = d(PE) + d(KE) + d(CE) + d(EE) + d(NE) + dU + \ldots \quad (4)$$

where the terms on the right stand, respectively, for potential, kinetic, chemical, electrical, nuclear, and internal energy. In an isolated system, $dE = 0$. With a little rearranging, we can rewrite Equation (4) to obtain:

$$dU = - (d(PE) + d(KE) + d(CE) + d(EE) + d(NE) + \ldots) \quad (5)$$

Equation (5) says that for an isolated system, any decrease in the sum of all the other forms of energy results in an increase in internal energy. But any increase in the internal energy of an isolated system, reflected, say, by a rise in temperature or a change of phase, is tantamount to an increase in the entropy of that system. As we concluded in the last chapter, $\Delta S = \Delta U/T$ for any constant-volume process at constant temperature, and $\Delta S = C_V \ln (T_2/T_1)$ for any constant-volume process during which the temperature changes.

ENTROPY AND THE SECOND LAW

In the last section, we took a cursory look at a wide variety of processes. Included were examples of the equalization of differences in temperature, pressure, and concentration as well as the transformation of kinetic, potential, electrical, chemical, and nuclear energy into internal energy. In each case, from our experience we knew, or were pretty sure of, the direction those processes would spontaneously take. We found that in every case the entropy of the system increased. Of course, our sampling was quite small, and it might seem risky to infer any universal truth for so few examples. Nevertheless, there is a pattern common to all the cases that seems entirely reasonable, so we might easily be persuaded to believe that *any* spontaneous

Each bit of spontaneity
Must generate some entropy.

process in an isolated system must result in an *increase* in the *entropy* of that system. This perhaps bold generalization has never been found wanting and is accepted as a universal principle known as the Second Law of Thermodynamics. As we have said, the Second Law can be stated in many ways whose equivalence is not always immediately obvious. For the coming discussion, we base our statement of the Second Law on the observations we have been making: *Any spontaneous process in any isolated system always results in an increase in the entropy of that system.*

There are some corollaries consequent to this particular statement of the Second Law:

1. If every spontaneous process in an isolated system results in an increase in entropy of that system, then for any such process to occur, an increase in the entropy of that system must be possible. If an entropy increase is impossible in a particular system, then no spontaneous change can occur in that system. The impossibility of spontaneous change must coincide with a maximum in the system entropy. This condition of maximum entropy, i.e., no possible spontaneous change, is known as the state of *absolute thermodynamic equilibrium.*

2. For any process it is always possible to define a system, that is, locate the boundary, in such a way that all parts of the real world affected in any way by the process are included within the system. On this basis, we conclude that any spontaneous process always results in an entropy increase. The ultimate extension of this notion includes the universe as a single system and becomes the basis for the Second Law statement that the entropy of the universe is always increasing. Extension to include the world led Clausius to his famous epigrammatic summary of the First and Second Laws: "Die Energie der Welt ist Konstant; die Entropie der Welt strebt ein Maximum zu" (The energy of the world is constant; the entropy of the world tends toward a maximum).

3. If every possible spontaneous change must result in an entropy increase, then the entropy of an isolated system can never decrease. Note that this statement refers only to an *isolated* system. It is always possible to decrease the entropy of a system by a suitable heat interaction, but it is not possible to do so by means of a work interaction. (To prove this last statement is a bit of an exercise in logic, but it is straightforward.)

4. Every imaginable process can occur, at least conceptually, in either the forward or reverse direction. The sequence of states in time can be in one order or its inverse. We have shown that any state spontaneously following on a previous state must have a higher entropy than the preceding

state. Thus, the sequence of all natural processes is determined by the principle of entropy increase. Therefore, the "direction of time" is itself determined by the entropy principle. For this reason, entropy has been called "Time's Arrow."

5. All of these conclusions are based on empirical observations that of necessity relate only to finite systems. If the universe is infinite, all bets are off and we cannot be sure that it is grinding to a halt. In fact, if the universe is finite but is many orders of magnitude larger than any isolated systems that we have been able to observe, it would be risky to assume that our statements of the thermodynamics laws apply.

Time and again in our discussions we have found that mathematical statements are much easier to manipulate and use than verbal statements. So it is appropriate to put our new statement of the Second Law into the terse and elegant language of mathematics. First, we note that if a system is isolated, its entropy must increase due to spontaneous changes within the system, or else nothing happens and the entropy remains unchanged. Thus:

$$d_iS \geq 0 \qquad (6)$$

where the subscript i refers to a change in entropy due to $i$rreversible processes $i$nside the system, $i$ndependent of heat interactions, that is in $i$solated systems. In sum, d_iS is any differential change in entropy uncompensated for by an equal and opposite entropy change in the surroundings. The Second Law says such changes must be equal to, or greater than, zero, never less. Included in d_iS are those entropy changes due to internal energy changes brought about by irreversible work interactions—as in stirring experiments. Such changes are actually due to production of internal energy inside the system. What crosses the system boundary is work.

We must recognize that the entropy of a system can be changed by a heat interaction with the surroundings. For this possibility we write:

$$d_eS = \delta q/T \qquad (7)$$

where the subscript e connotes an $e$xchange of entropy with the surroundings or a transfer to or from $e$xternal sinks or sources. As we have seen, the entropy of a system can be either increased or decreased by a heat exchange with the surroundings, depending upon whether δq is positive or negative.

The total differential change in the entropy of a system must be the sum of internal changes and exchanges with the surroundings. Thus:

$$dS = d_eS + d_iS \qquad (8)$$

and since $d_eS = \delta q/T$ by Equation (7),

$$dS = \delta q/T + d_iS \tag{9}$$

where always $$d_iS \geqq 0 \tag{10}$$

Equations (9) and (10) are a useful statement of the Second Law. Because d_iS is always positive, they can be combined to give:

$$dS \geqq \delta q/T \tag{11}$$

where the δq here is the heat exchanged with the surroundings. Only when the process is reversible, when $d_iS = 0$, can we write $dS = \delta q/T$. In the absence of heat interactions, that is to say for isolated or adiabatic systems, Equation (11) becomes

$$dS \geqq 0 \tag{12}$$

With the specified constraints either Equation (11) or (12), both of which were first asserted by Clausius, can be taken as statements of the Second Law. It seems more enlightening, however, if slightly less terse, to express the law in terms of how entropy can change (by internal *generation* or by *exchange* with the surroundings) as in Equations (8) through (10). The important substance of the Second Law is that the property called entropy of a system can be *produced* within a system (by irreversible processes) or can be *exchanged* with the surroundings as a consequence of heat interactions. Although entropy can be transferred back and forth between a system and its surroundings, it can never be destroyed. Once produced, it is forever on the debit side of the universal ledger.

Confusion arises when a system and its surroundings or another system have different temperatures. Let's examine an idealized situation characterized by an abrupt temperature change at the boundary between A and B with uniform but different temperatures in the regions on either side. We suppose that T_A is greater than T_B. For a small amount of heat interaction δq between them $d_eS_A = (\delta q/T)_A$ and $d_eS_B = (\delta q/T)_B$. The numerator δq's are the same size, but the sign for A is negative and for B positive. Thus, because T_A is larger than T_B, the amount of entropy d_eA transferred out of A is less than the amount of entropy d_eS_B transferred into B. This apparent discrepancy is discomfiting because it does not occur with conserved quantities like mass and energy with which we are generally more familiar and comfortable. Entropy is different because it can be created (but not destroyed!) In this case it is produced at the boundary where the temperature discontinuity occurs. If we wish, we can do the bookkeeping a little differently by drawing the boundary so that the temperature on either side of it is the same. Then the amount of entropy transferred into B would be the same as the amount transferred out of A, and the entropy production will occur inside either A or B—depending upon which side of the new boundary the temperature discontinuity is located. In this case, the entropy generation will show up as a d_iS term for either A or B.

In actuality, there is never a true temperature discontinuity at the boundary between a system and its surroundings (or another system). There is a gradient that is more or less steep depending upon the thermal conductivity of the material making up the system or surroundings. It is then expedient to consider entropy production as occurring in the gradient region where the temperature is changing with distance. In the case of a steady-state heat flow during which the temperature gradient remains constant, we must conclude that all of the properties of the gradient region including entropy are unchanging with time. More entropy flows out of this region than flows into it, and there emerges the concept of entropy production in a region associated with a flow of heat through it. Phenomena of this kind are the concern of a subject known as *Irreversible Thermodynamics*.

HOW ONE THING LEADS TO ANOTHER

This chapter began with a testimonial to the scope and importance of HER coupled with an apology for its humble origins. Having asserted that it embodied that fundamental principle, The Second Law of Thermodynamics, we decried its apparent preoccupation with the efficiency of machines. In what followed we arrived at, by inductive interpretation of a variety of observations, yet another general statement of the Second Law. It asserted that entropy is often created and never destroyed. It seems appropriate to ask whether HER can be derived from our new and more abstract statement and whether other famous statements may also be implicit in our assertions as to what entropy does in natural processes.

We begin by referring to the now familiar representation of a heat engine operating between reservoirs at T_1 and T_2 as shown in Figure 13-4. As the engine goes through a cycle, it absorbs an amount of heat Q_1 from the high-temperature reservoir, produces work W, and rejects an amount of heat Q_2 to the low-temperature reservoir at T_2. Because it has cycled, for the engine $\Delta S = 0$. Thus, by integrating Equation (11), we have

$$\oint \delta q/T = \frac{Q_1}{T_1} + \frac{Q_2}{T_2} \leq \Delta S = 0 \qquad (13)$$

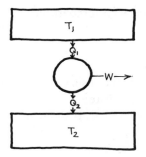

Figure 13-4 One last time: a heat engine operating between a source at T_1 and a sink at T_2.

or

$$Q_1/T_1 + Q_2/T_2 \leqq 0 \tag{14}$$

We multiply both sides of Relation (14) by T_2/Q_1 to obtain:

$$T_2/T_1 + Q_2/Q_1 \leqq 0 \tag{15}$$

We add one and subtract T_2/T_1 from both sides to get

$$1 + Q_2/Q_1 \leqq 1 - T_2/T_1 \tag{16}$$

that algebraically becomes

$$\frac{Q_2 + Q_1}{Q_1} \leqq \frac{T_1 - T_2}{T_1} \tag{17}$$

But by the First Law, $Q_2 + Q_1$ for a cycle of the engine is equal to the work done. Therefore, Equation (17) becomes

$$W/Q_1 \leqq \frac{T_1 - T_2}{T_1} \tag{18}$$

our old friend HER.

Now let's consider the possibility of running an engine continuously without rejecting any heat to a low-temperature reservoir. As we've already pointed out, this prospect is extremely attractive because there is an enormous amount of thermal energy in the oceans. If we could simply withdraw heat from the ocean into an engine without having to reject any waste heat, the energy crisis would be solved. This possibility would mean that a term Q_2/T_2 would not appear in the integrated form of Equation (9), and we would have for a system comprising high-temperature reservoir and engine:

$$\Delta S_{total} = Q_1/T_1 + \Delta_i S \tag{19}$$

Because Q_1 is negative, the entropy of the reservoir would be decreased. The remaining entropy change inside the system as we have defined it is represented by $\Delta_i S$ and must be associated with the engine. Clearly it is zero because the engine would have to cycle to run continuously. Thus, the net result of the overall process would be a decrease in entropy of the system comprising reservoir and engine. The system is adiabatic from an *overall* point of view because the only heat interaction occurs between the engine and the reservoir, inside the system. In short, withdrawing heat from a reservoir and converting it completely into work is equivalent to decreasing the entropy of an adiabatic or isolated system. According to our Statement of the Second Law, $d_i S \geqq 0$ (always), that cannot happen. The assertion that "it is impossible in a cyclic process to remove heat from a reservoir and convert it *completely* into work" is sometimes called the Kelvin–Planck statement of the Second Law.

Another famous statement of the Second Law, usually attributed to Clausius, says that "it is impossible to operate a cyclic process in such a way that the only effect will be to transfer heat from a reservoir at low temperature to another reservoir at higher temperature." This statement, assumed by Carnot in his cycle analysis and in keeping with Charlie's view of reality, also follows in a straightforward way from Equations (9) through (12).

THE ESSENCE OF ENTROPY

We started this chapter with the observation that the Zeroth Law asserts the existence of the property *temperature*. The First Law asserts the existence of *energy* as a property and declares that it can be neither created nor destroyed. After a long and perhaps painful introduction to the concept of *entropy* as a property we can now regard the Second Law as a statement that entropy can never be destroyed and is always being created by all kinds of natural processes. Let's hope that in the course of the discussion you will have come to some appreciation of how important and powerful the concept of entropy is. However, chances are that its essence still evades you. After all, what is it? To be sure, we have shown that it is a property, that its changes can be related to heat interactions by the expression $dS = \delta q/T$, and that in some way it is an inverse measure of the ability of an isolated system to change. Nevertheless, even after more exposure to it than you have here had, most people are nagged by the uneasy feeling that entropy is an abstraction that doesn't relate very directly with their everyday experience. You would probably have difficulty in explaining entropy, even in a crude sense, to someone you might meet on the street. Charlie would shake his head in dismay and trot off to look for supper. In the case of energy you could say, as you first learned in school, that it is "the ability to do work." Though not complete, exact, or entirely true, this statement gets across in a homely way an idea of what energy is all about. Charlie could relate to it. Is there some similar statement that can be made about entropy? Is there some way, even by analogy, to relate it to more familiar concepts?

Let's look again at what happened in some of the isolated systems that we examined earlier. When a gas was allowed to expand freely in a rigid, adiabatic vessel, we found that its entropy increased. Expansion from the same initial state to the same final state in a nonisolated system could have performed work on the surroundings. Because no work was performed during the free expansion, it would seem fair to say that the increase in entropy was accompanied by a loss of energy's ability to do work. When the expansion was carried out reversibly, allowing the piston to do work, the entropy of the gas increased by the same amount as in the free expansion. But in that case the increase in entropy of the gas was compensated for by a decrease in the entropy of an external reservoir. There was no net generation of new entropy. Entropy was simply transferred into the system from the surroundings. Thus, the generation of entropy during the work-less, free expansion corresponded to an unrecoverable loss of work.

In the case of the heat interaction between two blocks of copper initially at different temperatures, we found that there was also an entropy gain. In principle, we could have inserted a heat engine between the two blocks and performed work during the course of the temperature equalization. Thus, the entropy increase associated with the unresisted temperature equalization also represented a work loss. Similarly, in the course of equalizing the concentration of salt dissolved in water, work could have been performed. But the isolated system did no work when the diffusion was not resisted, and entropy was produced. That work potential was forever lost.

In the case of all those systems in which internal energy was generated at the expense of energy in other forms, there was also a loss of prospective work. The falling ball or the rotating flywheel could in principle have converted all their mechanical energy into external work. The internal energy that was formed, joule for joule and erg for erg, at the expense of the mechanical energy could only be partially converted into work. As we have found time and again, the fraction of internal energy the best heat engine can convert into work is determined by the temperature difference between

The higher does each index rise,
As more the world we "civilize."

source and sink and is always less than unity. Some of the internal energy withdrawn by a heat engine from a high-temperature source must always be rejected to a sink at lower temperature. In sum, the entropy generation associated with the transformation of other forms of energy into internal energy must also be regarded as an unrecoverable loss of ability to perform work.

Ruminating in this way, many people come to regard entropy as a measure of the extent to which energy is degraded, dissipated, or diluted so that it becomes less able to perform work. The energy of the world remains constant, but its usability diminishes with every increase in the world's entropy. Thus, entropy is an ever-increasing liability on the account books for energy transactions. Entropy does to energy what inflation does to currency— makes it ever less valuable. Thus, if energy is a measure of a system's ability to do work, entropy is a measure of how much this ability has depreciated or been devalued.

ORDER AND CHAOS

Another approach to the meaning of entropy comes out of a microscopic view of the real world, a view we have pretty much ignored. Though we have said very little about the structure of matter, we know that atoms and molecules are the stuff that make up the real world. A gram-mole of gas, remember, comprises 6.02×10^{23} molecules. Let's consider one such mole of gas in a rigid adiabatic container separated into equal volumes by a diaphragm with a hole in it. Let's suppose further that we christen one molecule a and baptize it with red paint to keep track of it. It will wander around colliding repeatedly with other molecules and with the container walls. In the long run, it will have equal access to all possible positions in the container. But its actual position at any time is determined by pure chance. The number

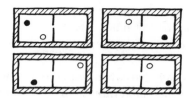

The molecules in chaos dance
To choreography by Chance.

of possible positions is the same on each side of the diaphragm because the volume on each side is the same. Therefore, we can expect on an average to find a on the left-hand side half of the time. In short, the probability that a is on the left is 1 in 2.

Now suppose we paint a second molecule white and call it b. By the same argument, we conclude that the probability that b will be on the left-hand side at any time is also 1 in 2. What is the probability that we will find both marked molecules on the left-hand side? Simply, $\frac{1}{2} \times \frac{1}{2} = (\frac{1}{2})^2 = \frac{1}{4}$. The chances are that half of the time a is on the left, b will also be on the left. Another way of looking at the problem is to note that there are four possible arrangements, $ab|$-$-$, a-$|b$, b-$|a$- and --$|ab$. Each of these possibilities is equally likely. Thus, the probability of any one is 1 in 4.

For three molecules, the probability that they will all be on the left-hand side is similarly $(\frac{1}{2})^3$, and the probability that n molecules will all be found on the left-hand side is $(\frac{1}{2})^n$. If n is only 20, the odds are already less than one chance in a million that all 20 molecules will be found on the left-hand side. But we have 6.02×10^{23} mol in the container. The chance that we will ever find all of them on the same side is a very small number indeed! Small, but nonetheless finite. It is not impossible for all the molecules to be on the left-hand side. It is just highly unlikely. (Don't bet any money that you will find them there, no matter what odds you can get.) On the other hand, the probability that we will find all the molecules in the container is unity as long as there are no leaks. Further, by a little reckoning you should be able to

Peas or missiles—'tis the same.
At any odds a no-win game!

show that, in general, no matter where the diaphragm is located, the probability of finding all the molecules on the left-hand side is $(V_1/V_2)^N$, where we understand that V_1 is volume to the left of the diaphragm and V_2 is the total volume of the cylinder.

In these terms, it seems fair to say that when we start out with all the molecules on the left-hand side and then puncture the diaphragm and let the gas expand to fill the container, we are allowing the gas to go from a state of low probability to a state of high probability. If we let P_1 equal this initial probability, we can say $P_1 = (V_1/V_2)^N$. As we have just pointed out, $P_2 = 1$. Because P_1 is likely to be such a small number, the easiest way to evaluate the ratio of the probabilities (P_2/P_1) is to subtract $\ln P_1$ from $\ln P_2$:

$$\ln P_1 = \ln(V_1/V_2)^N = N \ln(V_1/V_2) \tag{20}$$
$$\ln P_2 = 0 \tag{21}$$

Thus,

$$\ln P_2 - \ln P_1 = -N \ln (V_1/V_2) = N \ln (V_2/V_1) \tag{22}$$

We now multiply both sides of Equation (22) by R/N to obtain:

$$\frac{R}{N} \ln P_2 - \frac{R}{N} \ln P_1 = R \ln (V_2/V_1) \tag{23}$$

But as we saw in Equation (1), $R \ln V_2/V_1$ is the increase in entropy associated with free expansion of a gas from V_1 to V_2. Noting also that R/N is *Boltzmann's constant k*, or the "gas constant per molecule," we can write:

$$k \ln P_2 - k \ln P_1 = S_2 - S_1 \tag{24}$$

from which we infer:

$$S = k \ln P \tag{25}$$

where P represents the *thermodynamic probability* of the state for which the entropy is S.

Ludwig Boltzmann first realized this relation between entropy and probability. It is engraved on his tombstone in Vienna—with the slight difference that the thermodynamic probability is represented by the letter W instead of P. In these terms, the entropy of a system in a particular state is closely related to the probability of finding the system in that state. The increase in entropy during a spontaneous process corresponds to the transition from a state of lower probability to one of higher probability.

The simple analysis just performed relates the entropy of a population of molecules to the number of ways its members can distribute themselves in a volume to which they have access. The larger the volume, the greater the number of positions available to each molecule, the greater the number of ways they can be arranged, and the greater the entropy of the system comprising the population. In a real gas, any one molecule not only has a position, it has a velocity, and, therefore, a kinetic energy $(mv^2/2)$ that is the microscopic basis for what we have been calling internal energy for an ideal gas. Clearly, the greater the total internal energy of the population, the greater the number of options for energy content available to each molecule. The greater the number of such options, the more ways the system energy can be divided up among the molecules. Thus, by reasoning exactly analogous to that applied in the case of volume, it is possible to understand the increase in the entropy of a system as its internal energy is increased.

Closely related to these ideas is the notion that entropy increases with increasing disorder in a system. There seems to be a natural tendency for any system to go from order to chaos. I may start the week with my study in apple pie order, everything in its proper place. But unless I exert some effort to maintain that order, the chances are that by the end of the week there will have been a slide toward chaos. A deck of cards fresh from the manufacturer is so arranged that within each suit the cards ascend in order from ace to king. After a few shuffles, all semblance of this initial order disappears. Suppose you put a dollar's worth of pennies on the table, all with their heads up. If you sweep them on to the floor, the chances are pretty slim that they will come to rest either all heads or all tails. (In fact, the chance for that result is only $(\frac{1}{2})^{100}$ or about one in a billion billion!) The probability of disorder in a system is always higher than the probability of order.

This idea of relating entropy to disorder is very appealing. Consider the melting of ice, for example. On the basis of $dS = \delta q/T$, we found that the entropy of liquid water is higher than the entropy of ice at the same temperature. It is clear that molecules locked in place in an ice crystal are much more highly ordered than the same molecules moving randomly in liquid water. A valuable extension of these ideas leads to the concept of complete order that you might expect to find in a perfect crystal at a temperature of 0 K. In such a state, the molecules have no energy options available because the internal energy is zero. With respect to position they are also perfectly ordered if the crystal lattice is without defects. It would seem reasonable, therefore, to assign the value of zero to the entropy for this idealized state of maximum possible order (or minimum possible disorder). This assignment turns out to be very useful and is sometimes called the **Third Law of Thermodynamics.** It provides a basis for determining the absolute value of entropy for any system in any state.

But enough. How you accommodate to the concept of entropy is a matter of your own taste. It makes little difference whether you picture an increase in entropy as a loss of work, a depreciation of energy, an increase in probability, or a decrease in order. The important idea is that as the entropy of an isolated system increases, its ability to change decreases. When the entropy of such a system is at the maximum possible value consistent with its total energy, the system is in the state of absolute thermodynamic equilibrium. No further spontaneous change is possible.

Which sounds an appropriate final note. As a system ends its changing with a last increase in entropy, so must a book end its discourse with a last word. Let's hear it from Charlie.

Appendix I

Mechanical Properties—
Units and Measure

Systems we call *mechanical* are those whose structure and behavior can be described completely in terms of mass (m), length (l), and time (t). Force (f) is sometimes considered as a fourth dimension, but if we accept the law of motion that relates force to mass and acceleration, we can express force in terms of mass, length, and time. There are three systems of units widely used to measure these mechanical quantities: (1) The British Imperial System; (2) the Metric System; and (3) the Rational, or Giorgi, System—now internationally agreed upon and referred to as SI (for System Internationale). We will describe each of them.

TIME

Because astronomy was the earliest science to be concerned with the measurement of time, it is only natural that the unit of time common to all systems of mechanical units should be based on a characteristic astronomical interval. Thus, the **second** was for many years related to the earth's rotational period and defined as 1/86,400 of a mean solar day. At the 11th General Conference on Weights and Measures in 1960, it was even more precisely specified as 1/31,556,925.9747 of the tropical year 1900. In 1967 at the 13th General Conference on Weights and Measures, the second became officially defined as the "duration of 9,192,631,770 periods of the radiation corresponding to the transition between two hyperfine levels of the fundamental state of the atom of cesium 133." This specification was formulated so that the new second is experimentally indistinguishable from the "ephemeris second" based on the tropical year 1900. Thus, the fundamental unit of time is now defined as a large number of very small intervals rather than a very small fraction of a very large interval.

LENGTH

The Metric or cgs System   In this system, long the favorite of scientists even in English-speaking countries, c stands for centimeter, g for gram, and s for second. The unit of length is the **centimeter,** defined as $\frac{1}{100}$ of the standard **meter.** Originally, the meter was supposed to be $1/10^7$ of a meridional quadrant of the earth's circumference. Geodetic surveys along an arc from Dunkirk, France, to Montjuich near Barcelona in Spain were made, and a platinum bar was fashioned whose length was supposed to equal 1 ten-millionth of the quadrant distance. This bar was the standard meter for most of the nineteenth century. After the formation by treaty in 1875 of the International Bureau of Weights and Measures, the meter was redefined as the distance between two marks on a platinum–iridium bar of particular cross section stored at the International Bureau of Sèvres, a suburb of Paris. The earth's meridional quadrant is actually 1.00021×10^7 of those meters. Since the 1960 International Conference, the official meter is 1,650,763.73 wavelengths of the orange red line of krypton. Counting wavelengths of light by interferometry is a somewhat intricate business, but the new definition makes it possible for any laboratory to have access to the primary standard without having the expense (or pleasure) of a trip to Paris.

The SI or mks System   Also decimal, this m̲eter-k̲ilogram-s̲econd system is virtually the same as the c̲entimeter-g̲ram-s̲econd system. The numbers involved in calculations are identical. The difference is simply in the placement of the decimal point. The advantage of SI is that some of the derived units for work and force are of a more convenient magnitude and relate to electrical and magnetic units with less juggling of conversion factors. The fundamental source standards are the same. Thus, the official unit of length in SI is the meter as defined above.

The British Imperial or fps System   Long cherished by English-speaking engineers, the British system of units remains the practical standard in much of industry in the United States and the many other countries that once comprised the British Empire. Sooner or later it will be supplanted by the metric system (SI), which holds sway in most of the rest of the world. The big disadvantage of the British system is that it is not decimal, so calculations are more involved and errors are more likely. Nevertheless, it is so frequently encountered that we would be remiss if we did not use it in some of our examples.

The unit of length is the **foot,** which is subdivided into 12 **inches.** For a long time the foot was defined as one-third of the distance between two fine lines engraved on gold studs sunk in a specified bronze bar known as the "no. 1

standard yard'' cast in 1845. Since the 1960 International Conference, the inch is officially 2.54 centimeters and the foot is 12 such inches.

MASS

The Metric or cgs System   The unit of mass is the **gram,** defined as $\frac{1}{1000}$ of the mass of a platinum–iridium cylinder (whose length is equal to its diameter) stored at the International Bureau in Sevres. The mass of this cylinder was meant to be the same as the mass of 1000 cubic centimeters of water at 4°C, the temperature of maximum density. As it turned out, in light of more precise subsequent measurements, the volume of pure water having the mass of that reference cylinder, that is, a kilogram by definition, is 1,000.028 cm^3. For this reason, a secondary or derived unit of volume is the **liter** (L), originally defined as the volume occupied by 1 kilogram of pure water at 4°C. Most chemical glassware is calibrated in milliliters rather than cubic centimeters. One milliliter is equal to 1.000028 cubic centimeters, a difference hardly discernible in everyday titrimetry. Since 1964 the recommended definition of the liter has been 0.001 m^3 so the difference between the cubic centimeter and the milliliter has been defined away.

The SI or mks System   As in the case of the metric system, the source standard of mass is the **kilogram.** The difference is that the *unit* of mass is also the kilogram rather than the gram as in the cgs system.

The British or fps System   The practical unit of mass is the **pound** (lbm), defined as the mass of a cylinder of pure platinum about 1.35 inches high and 1.15 inches in diameter stored in London and sometimes called the avoirdupois pound. The pound is subdivided into 16 ounces or 256 drams or 7000 grains. Fourteen pounds equals 1 stone. (Eight stones equals 1 hundredweight, that is, 112 pounds!) The *rational* unit of mass in the British system is 32.174 pounds and is known as the **slug.** The reason for distinguishing between practical and rational will emerge when we discuss force. Meanwhile, we note that 1 lb is equivalent to 453.5924277 or approximately 454 g.

FORCE

We have noted that in addition to mass, length, and time, force is sometimes considered as a fourth quantity useful in describing mechanical systems. We also asserted that it was possible to describe force in terms of mass, length, and time by invoking Newton's laws of motion. Now we will enlarge on that assertion. First, we define the terms *velocity* and *acceleration*. By **velocity** we mean the rate at which position changes with time. It is the ratio of the distance traversed by an object in motion to the time required to travel that distance. Thus, velocity has the dimensions of distance/time, l/t, in appropriate units such as feet per second, miles per hour, centimeters per second, and so on. **Acceleration** is the rate at which velocity changes with time. It is the ratio of a change in velocity to the length of time during which the change occurred. Consequently, its dimensions are velocity/time = (distance/time)/time = distance/time^2, or l/t^2, for example, feet per second per second, or feet per second squared. Note that both velocity and acceleration are expressed in terms of length and time only.

Newton's First and Second laws of Motion can be combined to give:

$$f = ma$$

which says force is the product of mass and acceleration. But as we have seen, acceleration a can be expressed as l/t^2. Thus, we can write:

$$f = ma = ml/t^2$$

In other words, force can be expressed solely in terms of mass, length, and time. We will now examine the units for each of the three systems.

The Metric or cgs System   The unit of force is the **dyne.** A force of 1 dyne will accelerate a mass of 1 gram at a rate of 1 centimeter per second per second. Near the surface of the earth the average rate at which a falling object accelerates is about 980 centimeters per second per second. Thus, the force exerted by the earth's gravitational field on a mass of 1 gram is 980 dynes. Note that gravitational force decreases at increasing distances from the earth. Furthermore, as most of us in the space age are well aware, astronauts have frequently experienced situations in which the earth's gravitational field is just compensated by centrifugal force associated with orbital velocity and/or by an opposing gravitational attraction of the moon. Under these conditions of "weightlessness," there is no perceptible net force on any massive object, so it will float as though it were suspended motionless relative to nearby frames of reference such as the floor of the spacecraft. It is important to realize that such an object has lost its *weight* but not its *mass.* In other words **weight** is the gravitational force exerted on an object in proportion to its mass. Thus, the weight of an object depends upon where it is, but the *mass* of an object is independent of its position. Because the earth's gravitational attraction is so nearly uniform over the earth's surface, the *weight* of an object is almost universally used as a measure of its *mass*. The net result is that the terms are used interchangeably with much consequent confusion. For example, frequently one encounters the use of gram as a measure of force. What is meant is the force exerted by a mass of 1 gram in the earth's gravitational field, that is, its weight, actually 980 dynes.

The SI or mks System   The unit of force is the **newton (N)**. One newton will accelerate a mass of 1 kilogram at a rate of 1 meter per second per second. As we pointed out above, gravitational attraction near the earth's surface will accelerate an object at a rate of 980 cm or 9.8 m per second per second. Thus, the force exerted by gravity on a 1-kilogram mass near the earth's surface is 9.8 newtons. A newton is equal to 10^5 dynes.

The British or fps System   The unit of force is the **pound**! The exclamation point emphasizes the unfortunate use of the same word for both mass and force. To minimize confusion, the abbreviation lbf is often used for "pound force" and lbm for "pound mass." In our earlier discussion of mass units, we mentioned that the standard block of platinum in London was the *practical* unit of mass. A *rational* unit of mass would be that amount of mass that would be accelerated by unit force at a rate of unit distance per unit time per unit time. In the British system, that would be at a rate of 1 foot per second per second—1 ft/sec². As it happens, the unit of force is defined as the weight of (i.e., gravitational force exerted on) the reference block of platinum stored in London, the practical pound mass. But if that block of platinum were allowed to fall freely in the earth's gravitational field, it would accelerate at a rate of 32.174 ft/sec². Thus, the gravitational force that that block of platinum exerts in the earth's gravitational field would be sufficient to accelerate 32.174 times as much mass at a rate of 1 ft/sec². And so, the rational unit of mass is 32.174 pounds (lbm) and is known as the **slug**. One slug of mass will be accelerated at a rate of 1 ft/sec² by 1 pound of force (lbf). The force that would accelerate 1 actual *pound* of mass by 1 ft/sec² is thus 1/32.174 of 1 pound of force and is called a **poundal**. Confusing isn't it?

COMPOUND UNITS

There are innumerable ways in which the fundamental mechanical units of mass *m,* length *l,* and time *t* can be combined in ratios, products, and powers. Some of these combinations are so useful and are encountered so frequently that we assign them names of their own. The quantity force, which we discussed in the last section, is an example. It can be represented as the product of mass and acceleration, the latter being expressed as the ratio of length to the square of time, l/t^2. Thus, we say force has the dimensions of ml/t^2. Similarly, area has the dimensions l^2, volume the dimensions l^3, density the dimensions m/l^3 because it is mass per unit volume, and velocity the dimensions l/t. There are two quantities, whose units have compound dimensions, that we single out for further attention because they play such an important role in our discourse. They are work and pressure.

Work   The mechanical definition of **work** is in terms of the product of force and distance. Consequently, it has the dimensions ml^2/t^2. In the metric system the unit of work is given the name **erg.** It represents the amount of work done by a force of 1 dyne acting over a distance of 1 centimeter. The SI unit of work is the **joule.** It equals 1 newton-meter, 10^7 ergs. Because the erg is so tiny, the joule is often used as the unit of work in the metric system also. The British unit of work is called the foot-pound. As the name implies, it represents the amount of work done by a force of 1 pound acting over a distance of 1 foot.

   The rate at which work is performed is called **power.** Because it represents an amount of work divided by the time over which it is performed, that is, work/time, its dimensions are ml^2/t^3. The unit of power in both SI and the metric system is the **watt,** named after the Scot who saw how to save fuel by putting a separate condenser on the Newcomen steam engine. A watt is 1 joule per second, a relatively small rate of doing work. Kilowatts are much more commonly encountered in the rating of power-producing and power-consuming devices. In the British system, the unit of power is the **horsepower,** defined as 550 foot-pounds per second and equal to 0.745 kilowatts. Often you will find quantities of work being expressed as the product of power and time, for example, kilowatt-hours or horsepower-hours. This practice seems redundant because power is work divided by time. Why divide work by time and then multiply it by time in order to express it? We might just as well take the next step and express power in terms of kilowatt-hours/hours!

Pressure   We make great use of this property. It is force per unit area, or the ratio of force to the area over which it is distributed. My weight is about 160 lb. When I'm barefoot, my feet contact the floor over an area of about 0.4 ft² or 57.6 in² (rather a large area for my size because I happen to have flat feet). Thus, the pressure I exert (when I am standing) over that part of the floor with which my feet are in contact is 160/57.6 lb/in², or about 2.77 psi, where psi is a common abbreviation for "pounds per square inch."

Similarly, the tone arm on my record player is counterbalanced, so it is equivalent to about 2 g sitting on top of the stylus. The tip of the stylus is about 0.001 in (0.00254 cm) in diameter. If we assume that it is round and flat, the area in contact with the record is $\pi d^2/4$, or about $5 \times 10^{-6}\pi$ cm². Because 1 g exerts a downward force of 980 dynes, the pressure of the stylus on the record is 3.9×10^8 dynes/cm², which is equivalent to 5655 psi! No wonder a stylus tip must be made of diamond to keep from wearing out in a short time.

One dyne/cm² is a very small unit; so pressures will frequently be expressed in **bars** (metric unit). A bar is 10^6 dynes/cm² or 10^5 N/m², very close to a standard atmosphere—1.01325×10^5 N/m². One N/m² is a **pascal** (SI unit), in honor of the French scientist, Blaise Pascal. He was the first to demonstrate experimentally the principle of the barometer, which had been suggested by Descartes and Torricelli.

There are a number of other kinds of units having their roots in the nature of *hydrostatic pressure* and the ways in which it is measured. A characteristic feature of hydrostatic pressure is that it is *isotropic*—of equal magnitude in all directions. That is to say, any surface immersed in a fluid will experience the same pressure force, no matter what its orientation. A submerged submarine has almost the same force exerted on every square inch of its external surface whether that surface is vertical, horizontal, plane, or curved. (We say "almost" because the pressure exerted by the sea water depends upon the depth, and the bottom surfaces of the submarine are a bit further under water than the top surfaces.)

Let's consider ways of measuring hydrostatic pressure. We can imagine a piston in a cylinder as shown in Figure I-1. Assume that the cylinder is filled with gas and communicates with the fluid whose pressure is to be measured, perhaps in a tank somewhere. The upward force exerted on the piston is equal to the gas pressure multiplied by the area of the piston. The downward

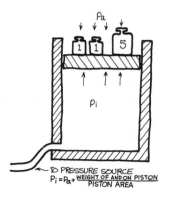

Figure I-1 Dead-weight pressure gauge.

force exerted by the piston equals the weight of the piston, plus any additional weights it may support, plus the pressure of the gas outside the cylinder multiplied by the area of the piston. (Note that it is the *projected* area that counts. The presence of weights or other surface irregularities makes no difference.) If the piston is perfectly frictionless it will move upward until we pile just enough weights on it to balance the upward and downward forces. Then the piston will remain stationary, and we can write:

$$p_i \times \text{Piston area} = p_a \times \text{Piston area} + \text{Total weight} \qquad (1)$$

where p_i is the gas pressure inside the cylinder, and p_a is the pressure of the gas in the ambient atmosphere. If we divide both sides by the piston area and rearrange terms, this equation becomes:

$$p_i - p_a = \text{Total weight/Piston area} \qquad (2)$$

Note that the dimensions are correct, that is, force/area. Also note that we are measuring the *difference* between p_i and p_a. If we want the absolute value of p_i, we must know the value of p_a. This feature is characteristic of almost all pressure gauges. They measure the difference between one pressure and another. To be sure, we can sometimes reduce the opposing pressure to zero, in this case, for example, by immersing the cylinder in a chamber at high vacuum where p_a would be negligibly small. Even so, the nature of the device is such that we would really be sensing the difference between one pressure and another that just happens to be negligibly small. The difference between p_i and p_a, that is, the left-hand side of the equation, is sometimes referred to as the *gauge* pressure, indicated by adding the letter g to the pressure dimensions. Thus, in the above equation if the total weight were 100 lb and the area of the piston 10 in^2, $(p_i - p_a)$ would be 10 psig. If p_a is equivalent to the standard atmosphere, it would be 14.7 psia where the last a indicates "absolute." In that case, the absolute value of p_i would be 24.7 psia.

The arrangement comprising a cylinder and piston with weights is actually used for measuring pressures and is sometimes referred to as a dead-weight gauge. Because it is difficult to obtain a good seal between the piston and the cylinder wall and at the same time keep the friction small, the dead-weight gauge is used mostly in the measurement of relatively high pressures—where the force required to overcome the frictional resistance of the piston is small relative to the total force on the piston exerted by the gas. A variation on this theme consists in balancing the pressure force on the piston by a calibrated

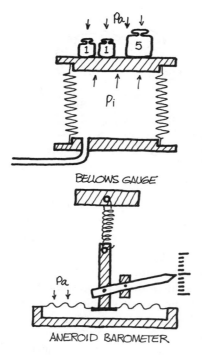

Figure I-2 Variations on the same theme: Modified dead-weight pressure gauges.

spring. Such an arrangement is the basis for the "pencil" tire-pressure gauge carried by many motorists and practically all service station attendants.

To overcome the difficulty of providing a seal between a moving piston and the cylinder wall, we can replace them with a cylindrical bellows having corrugated walls of thin metal that permit movement in the axial direction. The pressure force that tends to expand the bellows can be offset with weights or a spring until a balance is obtained. In another variation, the movable member can comprise a thin, corrugated diaphragm. The common *aneroid barometer* (aneroid is from the Greek and means literally without water) uses such a diaphragm to seal a shallow cavity that is evacuated. The atmospheric pressure bearing on the diaphragm is balanced by a spring as illustrated in Figure I-2. When the atmospheric pressure changes, the resulting motion of the diaphragm is magnified by a lever arrangement that moves a pointer over an indexed dial.

By far the most common of the devices not needing a moving seal is the **bourdon tube,** named after its French inventor, Eugene Bourdon. It is a metal tube of elliptical cross section bent into an arc as shown in Figure I-3. As the pressure in the tube increases relative to the outside pressure, the

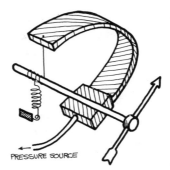

Figure I-3 Bourdon pressure gauge.

PRESSURE SOURCE

tube straightens out—in much the same way as the rolled-up paper tubes attached to a mouthpiece and used as party favors. The motion of the tip as the tube tries to straighten is connected by levers and gears to a pointer that moves over a dial on the face of the gauge. The thickness of the tube wall determines the flexibility of the tube and, therefore, the sensitivity of the gauge. Be reminded once more that all of these gauges sense the difference in pressure between the inside of the sensitive element and the ambient pressure. What they show, therefore, is a *gauge* pressure, $(p_i - p_a)$ in Equation (2).

We have saved till last what was undoubtedly the first of the many devices developed for pressure measurement. The liquid **manometer,** in spite of its age, is so widely used that its principles should be known by everyone. Its name comes from the Greek word *manos* meaning "gas" or "vapor," and applies generally to any instrument used for measuring the pressure of gases or vapors. Commonly, the term conjures up images of the liquid-in-tube devices that we will now examine.

Consider a flat-bottomed tube containing some liquid as shown in Figure I-4. The entire weight of the liquid column will be supported by the surface of the tube bottom. The magnitude of that force will equal the mass of the fluid

Figure I-4 Hydrostatic pressure: So-called even when the fluid is not water.

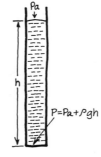

P_a

h

$P = P_a + \rho g h$

multiplied by the acceleration due to gravity, ρhAg where ρ is the liquid density, h the column height, A its cross-sectional area, and g the gravitational force. The resulting pressure at the bottom will be the total force divided by the area of the bottom, $\rho hAg/A = \rho hg = p$. Note that the pressure does not depend upon the cross-sectional area of the column. Because g is essentially the same everywhere on the earth's surface, the height of a column of specified liquid (which fixes ρ) becomes a direct indication of a pressure force. Thus, pressures are frequently reported in terms of inches, centimeters, feet, or millimeters of water (H_2O) or mercury (Hg), the two most widely used manometric liquids.

A pressure of 1 mm of mercury has been given the name **torr** in honor of Torricelli, the Italian physicist, a contemporary of Galileo who, along with the French scientist Descartes, is usually credited with conceiving the idea of the barometer. Actual reduction to practice is credited to Pascal. One torr equals 1333.22 dynes/cm². One ft of water equals 0.43352 psi, or 0.88265 in of mercury. Note that to be precise about the density of the liquid, one must also specify its temperature.

In the preceding discussion we considered the pressure force occasioned by gravitational attraction of the mass of the column of liquid. We did not take into account the force due to the pressure of the ambient atmosphere on the liquid surface at the top of the column. The absolute pressure on the bottom surface of the tube is the sum of the gas pressure plus the weight of the liquid column. Moreover, because the pressure in a fluid is isotropic—the same in all directions—the pressure on the wall of the tube at the bottom will be the same as the pressure on the bottom itself. (At any plane, real or imagined, that intersects the liquid column, the hydrostatic pressure is the same in all directions and equals the weight of the fluid column above that plane, plus the pressure exerted by any gas that may be in contact with the top surface of the liquid column.)

Now suppose there is another vertical tube containing a column of liquid of the same density and the same height. Suppose further that we provide a tube connecting the two liquid columns at their bottoms as in Figure I-5, so liquid can flow either way in response to any pressure difference that might exist. If both tubes are open to the same ambient atmosphere, the liquid columns will have the same height. The pressure at the bottom of each tube will be the same. We now connect the top of the right-hand tube to a container of gas at somewhat higher pressure than the ambient atmospheric pressure on the surface of the liquid column in the other tube. The pressure at the bottom of the right-hand tube will then be higher than the pressure on the bottom of the left-hand tube, and fluid will flow through the connecting tube until the pressure is the same at the bottom of each tube. As we have

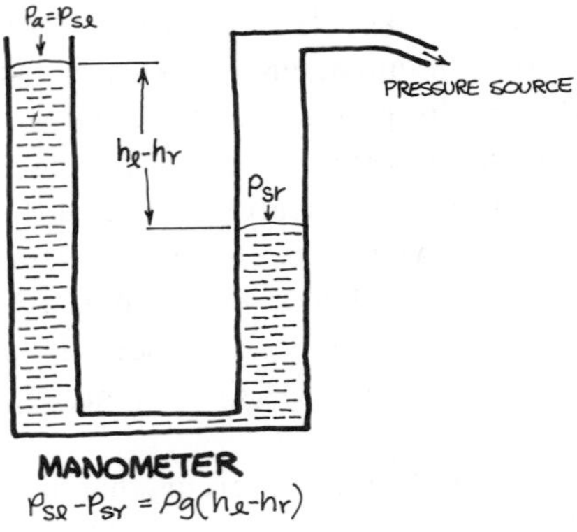

Figure 1-5 A simple manometer: The difference is what counts.

$P_a = P_{s\ell}$

$h_\ell - h_r$

P_{sr}

PRESSURE SOURCE

MANOMETER

$P_{s\ell} - P_{sr} = \rho g(h_\ell - h_r)$

already remarked, the bottom pressure in each tube equals the weight of the liquid column plus the gas pressure on the top surface.

Therefore, when the bottom pressures are equal, we can write:

$$p_{sr} + \rho g h_r = p_{sl} + \rho g h_l \qquad (3)$$

where subscript s refers to the surface of the liquid column, r to the right-hand tube, and l to the left. As before, ρ is the liquid density, g the gravitational force, and h the liquid height. A bit of algebraic manipulation on Equation (3) gives:

$$p_{sr} - p_{sl} = \rho g(h_l - h_r) \qquad (4)$$

In sum, the difference in height of two vertical columns of liquid connected at their bottoms is a measure of the difference in pressure at their surfaces. As in the case of the other gauges, manometers sense the difference between one pressure and another. If one of the legs of a manometer is open to the atmosphere, the difference in liquid levels corresponds to what we have called the gauge pressure.

We can connect one side of a manometer to a good vacuum pump and reduce the gas pressure to the lowest possible value, which will be the vapor pressure of the liquid. (We can't go below the vapor pressure because the liquid will continue to vaporize and replace all the vapor removed by the pump.) The differences in liquid levels will equal the absolute value of the atmospheric pressure (less the vapor pressure) if the other side of the mano-meter is open to the atmosphere. Such an arrangement is called a **barometer**. In actual barometers the evacuated side is sealed off, so the vacuum

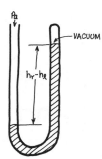

Figure I-6 The weatherman's right hand. One *atmosphere* is defined as $h_r - h_l = 760$ mm in a mercury barometer.

pump can be dispensed with as in Figure I-6. Because of its high density and because its vapor pressure is low enough to be negligible, about 10^{-3} mm Hg or torr, mercury is the liquid most often used in barometers. When the atmospheric pressure has its standard sea-level value, the height of the mercury column is exactly 760 mm, or 29.921 in. (The height of a water column that the atmosphere can support is 33.899 ft.)

Appendix II

A Guided Tour Through Log Land

WHAT ARE LOGARITHMS?

It is always possible to express any number in terms of another number raised to some power. If we let y represent any arbitrary number, we can write:

$$y = a^x \tag{1}$$

where x is the value of the *exponent*, or the *power*, to which we raise the number a in order to obtain the number y. In the language of *logarithms*, often called "logs" for short, x is called the "logarithm of y to the base a." In symbols:

$$\log_a y = x \tag{2}$$

If we choose the number 2 as the base, that is, let $a = 2$, and let $y = 4$, Equation (1) becomes:

$$4 = 2^2$$

In the logarithmic form of Equation (2):

$$\log_2 4 = 2$$

Similarly, when y is 8, x would be 3, or $\log_2 8 = 3$. If y is between 4 and 8, say 6, then x will be between 2 and 3, actually 2.59. Note that x is unity when $y = 2$. We call x for any value of y the "logarithm of y to the base 2" or the "log of y to the base 2." Thus, the log of 6 to the base 2 is 2.59.

A base more commonly used than 2 is 10. In this case,

$$y = 10^x \tag{3}$$

When y is 100, the value of x is 2, or "the log of 100 to the base 10 is 2." The log of 1000 to the base 10 is 3. The shorthand log form for these statements is:

$$\log_{10} 100 = 2$$
$$\log_{10} 1000 = 3 \tag{4}$$

Because they are so widely used, logarithms to the base 10 are known as *common logs*. In sum, the *logarithm* or *log* of any number is simply the exponent to which a particular base, commonly 10, must be raised in order to obtain that number.

USES OF LOGS

Logarithms are useful because they simplify computations. They change multiplications into additions and divisions into subtractions. This simplifying ability stems from their exponential origins. Suppose we want to multiply 8 by 4. In exponential form with 2 as a base, we can write:

$$8 \times 4 = 2^3 \times 2^2 = 2^{3+2} = 2^5 = 32 \tag{5}$$

In other words, to multiply numbers expressed in exponential form, add the exponents.

For the operations expressed in Equation (5), we can write in log form:

$$\log_2 8 = 3 \quad \log_2 4 = 2$$
$$\log_2 (8 \times 4) = \log_2 8 + \log_2 4 = 5 = \log_2 32 \tag{6}$$

Thus, to multiply two numbers, we add their logarithms. The result is the logarithm of the product. Of course, to take advantage of this simplification we need a means of determining the logarithm of a number or the number itself if we happen to have its logarithm. If we have a table of logs to the base 2, we can handle Equation (6) by looking up the logarithm of 8 (i.e., 3) and the logarithm of 4 (i.e., 2), add them to get 5, and then look in the table to find what number 5 is the logarithm of (32). In general, to carry out any multiplication, we simply find the log of each factor in the table and add them up.

The sum is the logarithm of the product. In the table, we find the number for which that sum is the logarithm. It is the product.

Division is reduced to subtraction. For example, in terms of common, or base 10, logarithms, the log of 1000 is 3 and the log of 100 is 2. We wish to divide 1000 by 100. We can write in exponential form:

$$1000/100 = 10^3/10^2 = 10^{(3-2)} = 10^1 = 10 \tag{7}$$

In log language, we would write:

$$\log_{10}(1000/100) = \log_{10}1000 - \log_{10}100$$
$$= 3 - 2 = 1 = \log_{10}10 \tag{8}$$

In general, to divide one number by another, we subtract the logarithm of the divisor from the logarithm of the dividend. The difference is the logarithm of the quotient. Thus, the log of any fraction is the log of the numerator less the log of the denominator. One dividend from learning how to divide by logs is that we can now show that the log of unity is zero—that any number raised to the zero power is unity. Let's divide 1000 by itself:

$$\log_{10}(1000/1000) = \log_{10}1000 - \log_{10}1000$$
$$\log_{10}1 = 3 - 3 = 0 \tag{9}$$

We obtain the same result, no matter what base is used.

It is easy to extend the operations of multiplication and division to the raising to powers and extracting of roots. Let's consider raising 10 to the fifth power. In exponential form:

$$10^5 = 10 \times 10 \times 10 \times 10 \times 10 = 10^{(1+1+1+1+1)}$$
$$= 10^{(5\times1)} \tag{10}$$

In log form this operation becomes:

$$\log_{10}10^5 = \log_{10}(10 \times 10 \times 10 \times 10 \times 10)$$
$$\log_{10}10^5 = \log_{10}10 + \log_{10}10 + \log_{10}10 + \log_{10}10$$
$$= 5 \log_{10}10 \tag{11}$$

In words, the logarithm of a number raised to a power is simply that power multiplied by the logarithm of the number, or in general:

$$\log y^n = n \log y \tag{12}$$

Extracting a root is tantamount to raising to a fractional power. Thus:

$$10 = (100)^{1/2} = (10^2)^{1/2}$$
$$= 10^{(2 \times 1/2)} = 10^1 = 10 \tag{13}$$

In log form:

$$\log_{10} 10 = \log_{10}(100)^{1/2} = \log_{10}(10^2)^{1/2}$$
$$= (\tfrac{1}{2}) \log_{10}(10^2)$$
$$= (\tfrac{1}{2}) \times 2 = 1 \tag{14}$$

In general terms, we can write:

$$\log y^{1/n} = (1/n) \log y \tag{15}$$

which is true for any base.

We can summarize the basis of computation by logarithms as follows:

$$\log m + \log n = \log m \times n \tag{16}$$

$$\log m - \log n = \log m/n \tag{17}$$

$$\log m^r \qquad = r \log m \tag{18}$$

Of course, the examples that we have used would not suggest that logarithmic notation offers any particular attractions. To multiply 7 by 5, we would hardly bother to add the logarithm of 7 to the logarithm of 5 and find out what number has the resulting sum as its logarithm. But in more intricate calculations, logs can be quite helpful. Consider the following operation:

$$\frac{459 \times 764 \times 346}{223 \times 65 \times 127}$$

To carry out all these multiplications and divisions longhand would take quite a while. In logarithmic operations, we would merely add the logs of the terms in the numerator, subtract the sum of the logs in the denominator, and find the logarithm of the answer. Of course, we need a table to determine the logs, and the accuracy of the calculation would depend on the precision of the table, that is, the number of places to which the logarithms are expressed. The four-place table that appears on pp. 274–275 is adequate only for results with three significant figures. Let's carry out a few operations as practice in using the table. But first, a few preliminary remarks.

The table is of common logarithms to base 10. In such a decimal system, the logarithm of any number—the exponent to which 10 must be raised to give that number—will generally have a decimal, or nonintegral, value. For example, the log of 10 is 1.0000. The log of 100 is 2.0000. Any number between 10 and 100 will have a logarithm whose value is between 1.0000 and 2.000. Similarly, any number between 100 and 1000 will have a logarithm whose value is between 2.0000 and 3.0000. Consider the number 546. Clearly

$$\log 546 = 2.xxxx \tag{19}$$

Similarly,

$$\log 54.6 = 1.xxxx \tag{20}$$

But we also note that

$$\log 546 - \log 54.6 = \log 546/54.6 = \log 10 = 1.0000$$

and

$$= 2.xxxx - 1.xxxx = 1.0000 \tag{21}$$

The only way this can happen is if $xxxx$ is the same in both cases. In other words, the part of the logarithm to the right of the decimal point will be the same for any number comprising the digits 546. This part of the logarithm, $xxxx$, is known as the **mantissa**. The number to the left of the decimal in the logarithm determines where the decimal point is located; it is known as the **characteristic** of the logarithm.

What we get out of a log table is the *mantissa*, the numbers in the logarithm to the right of the decimal point. If you look on the bottom line of the first part of the table, you will find all 3-digit numbers having 54 as the first 2 digits. The first column corresponds to 540, the fifth to 544, and so on. We are interested in 546. Its mantissa (in the seventh column) is 7372. Thus:

$$
\begin{aligned}
\log 546 &= 2.7372 \\
\log 54.6 &= 1.7372 \\
\log 5.46 &= 0.7372 \\
\log 5{,}460{,}000 &= 6.7372
\end{aligned} \tag{22}
$$

What happens in the case of 0.546? Well, we can write:

$$0.546 = 5.46/10 \tag{23}$$

so $$\log 0.546 = \log 5.46 - \log 10 \qquad (24)$$

$$= 0.7372 - 1.0000 = -0.2628 \qquad (25)$$

Often, instead of carrying out the subtraction, we write

$$\log 0.546 = \bar{1}.7372 \qquad (26)$$

where the minus sign above the characteristic indicates that it is negative. The mantissa is always positive unless there is a colinear minus sign in front as in Equation (25).

Let's carry out a sample calculation. Let's multiply 345 by 567 and divide the result by 891, or $(345 \times 567)/891$. In log language, this becomes:

$$\log 345 + \log 567 - \log 891 = \log (345 \times 567)/891 \qquad (27)$$

We get the mantissas from the table. The characteristic in each case is 2. Therefore, Equation (27) becomes:

$$2.5378 + 2.7536 - 2.9499 = 2.3415 \qquad (28)$$

The mantissa nearest 3415 from below is 3404, corresponding to 219. The next highest value is 3424, corresponding to 220. Because 3415 is half way between 3404 and 3424, we conclude that the number for which 2.3415 is the logarithm would be half way between 219 and 220 or 219.5, the best answer we can get from a four place log table.

Note that the process of *interpolation* by which we chose the value between 219 and 220 has allowed us to obtain one more significant figure. If you carry out the above operation longhand, or on a calculator, you will get the value 219.54545 if you go to eight significant figures. In these days of pocket calculators and desk computers, log tables are not used nearly as much as they once were. Nevertheless, it will pay you to carry out a few operations with the accompanying table to get the feel of how logs work. Moreover, for reasons about to emerge, it is often useful to know what the logarithm of a number is per se.

LOGARITHMS AND VERY SMALL NUMBERS

Let's now go back to exponential notation and consider what happens when we have numbers near unity. If $y = a^x$, we have seen that y is unity when x is zero. Clearly, therefore, if x is a very small number, y will be only slightly larger than unity. The smaller x becomes, the nearer unity will be y. We naturally ask if other relationships between numbers have a similar property. There is one very simple one:

$$y = 1 + x \tag{29}$$

In this case too as x approaches zero y will approach unity. When x is small, y will be a number only slightly larger than unity. Therefore, let us arbitrarily write:

$$a^x = 1 + x \tag{30}$$

Because both sides of this equation approach unity as x goes to zero, what particular value of a will best satisfy the equation for small values of x? We solve for a by taking the xth root of both sides to get:

$$a = (1 + x)^{1/x} \tag{31}$$

We now ask what value the right-hand side approaches as x gets very small. Quickly we find x cannot be equal zero on the right-hand side because that is unity raised to the infinity power—which we don't know how to evaluate. Instead, we construct a table of values for a over a range of small values for x on either side of zero.

x	$(1 + x)^{1/x} =$	a
1.0	2^1	2.00
0.1	1.1^{10}	2.59
0.01	1.01^{100}	2.70
−0.01	0.99^{-100}	2.73
−0.1	0.9^{-10}	2.87

Clearly, when x is zero the value of a will be between 2.70 and 2.73. Extensive computations show that the limiting value of $(1 + x)^{1/x}$ as x approaches zero is 2.71828183 .... This number, like π, the ratio of the circumference of a circle to its diameter, is a never-ending decimal. It is given the symbol e and is the base of *natural* logarithms, sometimes called *Napierian* logarithms in honor of John Napier, the Scottish mathematician who invented them. Because of the way we have arrived at its value, e is the best choice for base a in Equation (30). Consequently, we can write:

$$e^x = 1 + x \qquad (32)$$

and be confident that it is a reasonably accurate expression for small values of x. It becomes exactly correct in the limit as x approaches zero. In log form, Equation (32) becomes:

$$\log_e(1 + x) = x = \ln(1 + x) \qquad (33)$$

The last term on the right points out that ln is the customary shorthand for $\log_e$, an acronym for *log natural*. Equation (33) is the most important result, for our present purpose, of this excursion into log land. It says that we can substitute for any small number the natural log of 1 plus the number, and vice versa. This substitution is most useful as you will learn in Chapter 4.

For general computations, natural logs are not nearly as convenient as common logs. Therefore, tables of natural logs are not as common. But it is easy to obtain the natural log from the common log. Recall:

$$\text{if} \quad y = e^x \quad \text{then} \quad \ln y = x$$

and
$$\log y = x \log e = (\ln y)(\log e)$$

Therefore,
$$\ln y = (1/\log e)(\log y)$$
$$= 2.303 \log y \qquad (34)$$

You might amuse yourself by verifying this result with the log table.

A table of four-place logarithms follows on pages 274–275.

Four-Place Logarithms

N	0	1	2	3	4	5	6	7	8	9
10	0000	0043	0086	0128	0170	0212	0253	0294	0334	0374
11	0414	0453	0492	0531	0569	0607	0645	0682	0719	0755
12	0792	0828	0864	0899	0934	0969	1004	1038	1072	1106
13	1139	1173	1206	1239	1271	1303	1335	1367	1399	1430
14	1461	1492	1523	1553	1584	1614	1644	1673	1703	1732
15	1761	1790	1818	1847	1875	1903	1931	1959	1987	2014
16	2041	2068	2095	2122	2148	2175	2201	2227	2253	2279
17	2304	2330	2355	2380	2405	2430	2455	2480	2504	2529
18	2533	2577	2601	2625	2648	2672	2695	2718	2742	2765
19	2788	2810	2833	2856	2878	2900	2923	2945	2967	2989
20	3010	3032	3054	3075	3096	3118	3139	3160	3181	3201
21	3222	3243	3263	3284	3304	3324	3345	3365	3385	3404
22	3424	3444	3464	3483	3502	3522	3541	3560	3579	3598
23	3617	3636	3655	3674	3692	3711	3729	3747	3766	3784
24	3802	3820	3838	3856	3874	3892	3909	3927	3945	3962
25	3979	3997	4014	4031	4048	4065	4082	4099	4116	4133
26	4150	4166	4183	4200	4216	4232	4249	4265	4281	4298
27	4314	4330	4346	4362	4378	4393	4409	4425	4440	4456
28	4472	4487	4502	4518	4533	4548	4564	4579	4594	4609
29	4624	4639	4654	4669	4683	4698	4713	4728	4742	4757
30	4771	4786	4800	4814	4829	4843	4857	4871	4886	4900
31	4914	4928	4942	4955	4969	4983	4997	5011	5024	5038
32	5051	5065	5079	5092	5105	5119	5132	5145	5159	5172
33	5185	5198	5211	5224	5237	5250	5263	5276	5289	5302
34	5315	5328	5340	5353	5366	5378	5391	5403	5416	5428
35	5441	5453	5465	5478	5490	5502	5514	5527	5539	5551
36	5563	5575	5587	5599	5611	5623	5635	5647	5658	5670
37	5682	5694	5705	5717	5729	5740	5752	5763	5775	5786
38	5798	5809	5821	5832	5843	5855	5866	5877	5888	5899
39	5911	5922	5933	5944	5955	5966	5977	5988	5999	6010
40	6021	6031	6042	6053	6064	6075	6085	6096	6107	6117
41	6128	6138	6149	6160	6170	6180	6191	6201	6212	6222
42	6232	6243	6253	6263	6274	6284	6294	6304	6314	6325
43	6335	6345	6355	6365	6375	6385	6395	6405	6415	6425
44	6435	6444	6454	6464	6474	6484	6493	6503	6513	6522
45	6532	6542	6551	6561	6571	6580	6590	6599	6609	6618
46	6628	6637	6646	6656	6665	6675	6684	6693	6702	6712
47	6721	6730	6739	6749	6758	6767	6776	6785	6794	6803
48	6812	6821	6830	6839	6848	6857	6866	6875	6884	6893
49	6902	6911	6920	6928	6937	6946	6955	6964	6972	6981
50	6990	6998	7007	7016	7024	7033	7042	7050	7059	7067
51	7076	7084	7093	7101	7110	7118	7126	7135	7143	7152
52	7160	7168	7177	7185	7193	7202	7210	7218	7226	7235
53	7243	7251	7259	7267	7275	7284	7292	7300	7308	7316
54	7324	7332	7340	7348	7356	7364	7372	7380	7388	7396

Four-Place Logarithms (*continued*)

N	0	1	2	3	4	5	6	7	8	9
55	7404	7412	7419	7427	7435	7443	7451	7459	7466	7474
56	7482	7490	7497	7505	7513	7520	7528	7536	7543	7551
57	7559	7566	7574	7582	7589	7597	7604	7612	7619	7627
58	7634	7642	7649	7657	7664	7672	7679	7686	7694	7701
59	7709	7716	7723	7731	7738	7745	7752	7760	7767	7774
60	7782	7789	7796	7803	7810	7818	7825	7832	7839	7846
61	7853	7860	7868	7875	7882	7889	7896	7903	7910	7917
62	7924	7931	7938	7945	7952	7959	7966	7973	7980	7987
63	7993	8000	8007	8014	8021	8028	8035	8041	8048	8055
64	8062	8069	8075	8082	8089	8096	8102	8109	8116	8122
65	8129	8136	8142	8149	8156	8162	8169	8176	8182	8189
66	8195	8202	8209	8215	8222	8228	8235	8241	8248	8254
67	8261	8267	8274	8280	8287	8293	8299	8306	8312	8319
68	8325	8331	8338	8344	8351	8357	8363	8370	8376	8382
69	8388	8395	8401	8407	8414	8420	8426	8432	8439	8445
70	8451	8457	8463	8470	8476	8482	8488	8494	8500	8506
71	8513	8519	8525	8531	8537	8543	8549	8555	8561	8567
72	8573	8579	8585	8591	8597	8603	8609	8615	8621	8627
73	8633	8639	8645	8651	8657	8663	8669	8675	8681	8686
74	8692	8698	8704	8710	8716	8722	8727	8733	8739	8745
75	8751	8756	8762	8768	8774	8779	8785	8791	8797	8802
76	8808	8814	8820	8825	8831	8837	8842	8848	8854	8859
77	8865	8871	8876	8882	8887	8893	8899	8904	8910	8915
78	8921	8927	8932	8938	8943	8949	8954	8960	8965	8971
79	8976	8982	8987	8993	8998	9004	9009	9015	9020	9025
80	9031	9036	9042	9047	9053	9058	9063	9069	9074	9079
81	9085	9090	9096	9101	9106	9112	9117	9122	9128	9133
82	9138	9143	9149	9154	9159	9165	9170	9175	9180	9186
83	9191	9196	9201	9206	9212	9217	9222	9227	9232	9238
84	9243	9248	9253	9258	9263	9269	9274	9279	9284	9289
85	9294	9299	9304	9309	9315	9320	9325	9330	9335	9340
86	9345	9350	9355	9360	9365	9370	9375	9380	9385	9390
87	9395	9400	9405	9410	9415	9420	9425	9430	9435	9440
88	9445	9450	9455	9460	9465	9469	9474	9479	9484	9489
89	9494	9499	9504	9509	9513	9518	9523	9528	9533	9538
90	9542	9547	9552	9557	9562	9566	9571	9576	9581	9586
91	9590	9595	9600	9605	9609	9614	9619	9624	9628	9633
92	9638	9643	9647	9652	9657	9661	9666	9671	9675	9680
93	9685	9689	9694	9699	9703	9708	9713	9717	9722	9727
94	9731	9736	9741	9745	9750	9754	9759	9763	9768	9773
95	9777	9782	9786	9791	9795	9800	9805	9809	9814	9818
96	9823	9827	9832	9836	9841	9845	9850	9854	9859	9863
97	9868	9872	9877	9881	9886	9890	9894	9899	9903	9908
98	9912	9917	9921	9926	9930	9934	9939	9943	9948	9952
99	9956	9961	9965	9969	9974	9978	9983	9987	9991	9996

After R. N. Smith and C. Pierce, *Solving General Chemistry Problems*, 5th ed. Copyright © 1980 by W. H. Freeman and Company.

Appendix III

Everything Has Entropy

In Chapter 12, we found that for systems obeying the equation of state, $pV = nRT$, we could write $dS = \delta q/T$ where T is the absolute temperature (Kelvin or Rankine) of the system during a small amount of heat interaction δq between the system and its surroundings (or another system). This expression presumes a *uniform temperature throughout* the system. Otherwise, we could not assign a value to T. In turn, temperature uniformity implies that the rate of heat interaction δq must be very slow, that is to say, *reversible* from the standpoint of the system (but not necessarily reversible from the standpoint of the surroundings).

In effect, the relation $dS = \delta q/T$ asserts the existence of a property, entropy S, whose differential dS is exact (or perfect or complete or total). A useful feature of quantities that are properties and whose differentials are exact is that the integral sums of such differentials are simply the changes in their value between one system state and another, and are independent of the path the system takes from the initial to the final state. The integral sums of *inexact* differentials depend very much on the path. Recall that the First Law, $dE = \delta q - \delta w$, similarly asserts the existence of the property, energy E, and defines an exact differential change dE in terms of the difference between the inexact differentials δq for heat and δw for work. The existence of energy and the exactness of its differential proved to be exceedingly useful in our applications of the First Law to the thermodynamic analysis of various problems. The existence of entropy and the exactness of its differential are also extremely useful in thermodynamic analysis.

In Chapter 12, we asserted that entropy exists and can be characterized by $dS = \delta q/T$ for *any* system—not just those obeying $pV = nRT$. How do we justify this assertion? First, to refresh our memories, let's review the argument that showed $dS = \delta q/T$ was true in the case of an ideal gas. For 1 mol of such a gas at rest in its frame of reference, the First Law can be written:

$$\delta q = dU + \delta w$$

$$= C_V \, dT + p \, dV$$

$$= C_V \, dT + RT \, dV/V \tag{1}$$

where dU is a differential change in internal energy, and C_V is the constant-volume molar heat capacity. We divide both sides of Equation (1) by T to get:

$$\delta q/T = C_V \, dT/T + R \, dV/V$$

Integrating from initial state 1 to final state 2 yields:

$$\int_1^2 \delta q/T = C_V \ln (T_2/T_1) + R \ln (V_2/V_1) \tag{2}$$

The value of $\int_1^2 \delta q/T$ is determined only by the values of T and V in the initial and final states. Consequently, it does not depend upon the path or particular sequence of states the system follows from state 1 to state 2. Thus, $\delta q/T$ is an exact differential dS of a quantity having all the characteristics of a property. We call it **entropy** and represent it by the symbol S. Consequently, we can write:

$$\int_1^2 \delta q/T = \int_1^2 dS = S_2 - S_1$$

Equivalent evidence of the exactness of $\delta q/T$ is contained in the observation that in a cyclic process, for which the initial and final states are the same, both terms on the right-hand side of Equation (2) vanish so that

$$\oint \delta q/T = \oint dS = 0$$

If the value of an integral is zero for a cyclic process, the integrand, in this case $\delta q/T$, is an exact differential, in this case dS.

We now consider the behavior of an arbitrary System A in an arbitrary process. We require that during the course of this arbitrary process, any heat and work interactions System A undergoes are with a System I comprising an ideal gas. This requirement is not a constraint upon System A because no system is concerned with the nature of what is on the other side of its boundary. If System A does work, it would just as soon compress a gas as raise a weight or charge a battery. The *direct* consequence outside the system of any work interaction might take on a variety of forms, but as we learned in Chapter 1, any of these forms can be reduced to raising a weight or compressing a gas. Similarly, in the case of a heat interaction, the nature of the partner in the heat exchange is of no consequence. All that the system knows or cares about is whether it gains or loses energy by a heat transfer. It is oblivious to the nature of systems beyond its boundary.

We impose one further condition. We require that the temperature of System I and System A always be the same. This condition is easily met. We simply have a thermometer in System A to indicate its temperature at all times. We allow reversible adiabatic work interactions between System I and a third system, System 3, that can release, absorb, and store mechanical energy—for example, a spring or an arrangement of weights and pulleys. Thus, when the temperature of System A tries to go higher than that of System I, we call on System 3 to compress System I adiabatically and raise its temperature. Similarly, when the temperature of System A tries to go below that of System I, we simply let the latter do adiabatic work on System 3 and cool off. In other words, by means of reversible, adiabatic work interactions between System I and System 3, we can maintain identical temperatures of System I and System A throughout any process.

The situation just described is represented schematically in Figure III-1. The subscripts A, I, and 3 refer, respectively, to arbitrary System A, ideal gas System I, and the reservoir of mechanical energy, System 3. The order of the subscripts indicates from which system the interaction is viewed. Thus, δw_{I3} indicates that System I has either gained or lost energy by means of a work interaction with System 3. The symbol δq_{AI} means that System A has either gained or lost energy by means of a heat interaction with System I. Clearly, an interaction viewed by one system is equal in magnitude but opposite in sign to the same interaction as viewed by its partner in that interaction. Thus $\delta q_{AI} = -\delta q_{IA}$.

Let's allow System A to go through an arbitrary cyclic process for which, of course, the initial and final states must be identical. For such a process, the First Law says that $\oint dE = 0$. Because $dE = \delta q - \delta w$:

$$\oint \delta q_{AI} - \oint \delta w_{AI} = 0$$

But each $\delta q_{AI} = -\delta q_{IA}$, and each $\delta w_{AI} = -\delta w_{IA}$. Therefore, from the standpoint of System I:

$$\oint - \delta q_{IA} - \oint - \delta w_{IA} = 0 = \oint \delta q_{IA} - \oint \delta w_{IA}$$

ARGUMENT

1. BECAUSE $T_1 = T_A$ AT ALL TIMES, AT THE END OF SYSTEM A'S CYCLE: $\Delta T_A = \Delta T_1 = \Delta U_1 = \Delta E_1 = 0$

2. BUT $\Delta E_A = \Delta E_{AI} = 0 = \Delta E_1$. THEREFORE, $\Delta E_{31} = \Delta E_1 = 0$, AND SYSTEM 3 HAS CYCLED.

3. IF SYSTEMS 3 AND A HAVE CYCLED, ALL COORDINATES AND PROPERTIES HAVE INITIAL VALUES AND NO DISPLACEMENT OF SYSTEM I HAS OCCURRED. THEREFORE $\Delta V_1 = \Delta T_1 = \Delta U_1 = 0$ AND SYSTEM I HAS CYCLED.

4. BUT FOR A CYCLIC PROCESS IN AN IDEAL GAS $\oint q_{1A}/T = 0$

5. BECAUSE EACH $\delta q_{1A} = -\delta q_{AI}$ AND $T_A = T_1$ ALWAYS:
$$\oint -\delta q_{AI}/T = 0 = \oint \delta q_{AI}/T$$

6. CONSEQUENTLY, $\delta q_{AI}/T = dS_A$ IS AN EXACT DIFFERENTIAL, Q.E.D.

which means that there has been no energy transferred from System A to System I. Moreover, because $T_I = T_A$ at all times and because T_A is at its initial value after the cycle, T_I must also be at its initial value. The internal energy of an ideal gas depends only upon its temperature. Therefore, the energy of System I will return to its initial value when System A has cycled. Consequently, there can have been no net transfer of energy from System 3 to System I, and System 3's energy will return to its initial value.

Because there have been only work interactions between System 3 and System I, the return of System 3's energy to its initial value means that it can have performed no net compression work on System I and, therefore, brought about no change in System I's volume. But System A, having cycled, has also returned to its initial state. Its volume and all its position coordinates must have their initial values. It follows that System A cannot have changed the volume in System I. Thus, T_I and V_I both have their initial values, so System I has cycled. The net result of the arbitrary, cyclic process in System A has been a cyclic process in both System I and System 3 as well.

Throughout their cycles, the temperatures of Systems A and I have been the same at all times, so each $\delta q_{1A}/T$ has equaled its corresponding $-\delta q_{AI}/T$. As we have already shown, for an ideal gas, $\oint \delta q/T = 0$ in any cycle. Thus:

$$\oint q_{1A}/T = \oint -\delta q_{AI}/T = 0 = \oint \delta q_{AI}/T$$

Consequently, $\delta q_{AI}/T$ must also be an exact differential. But, as we have already noted, System A is unconcerned about the particular partner with which it may be having a heat or work interaction. Therefore, for *any* arbitrary system in *any* reversible heat exchange (reversible in the sense that the temperature of the system must be uniform at all times to specify T):

$$\delta q_A/T = dS_A$$

Because there have been no constraints on the structure, or composition, of System A, it follows that the property entropy can be characterized and its changes evaluated for any system whatsoever—Q.E.D.

Though simple, the argument you have just endured has been intricate and somewhat involved. To bring it into qualitative perspective here is a brief recapitulation:

1. We let an arbitrary system go through an arbitrary cyclic process in which all of its heat and work interactions were with a collaborating system comprising an ideal gas.

2. By means of adiabatic reversible work interactions with a third system the temperature of the ideal gas system was kept equal to the temperature of the arbitrary system at all times. Thus, at the end of the arbitrary system's cycle the ideal gas system had also cycled because its properties including temperature, and therefore energy, had returned to their initial values.

3. When an ideal gas system cycles, $\oint \delta q/T = 0$ so $\delta q/T$ must be the exact differential of a property. Because every $\delta q/T$ for the ideal gas system during its cycle was equal to $-\delta q/T$ for the arbitrary system, $\oint \delta q/T = 0$ for the arbitrary system also. Therefore, $\delta d/T$ is exact for any system and can be taken as the differential of the property we call entropy, that is to say dS.

Appendix IV

The Atomic Weights
of The Elements

The Atomic Weights of the Elements*

Element	Symbol	Atomic Number	Atomic Weight	Element	Symbol	Atomic Number	Atomic Weight
Actinium	Ac	89	[227]	Europium	Eu	63	151.96
Aluminum	Al	13	26.9815	Fermium	Fm	100	[257]
Americium	Am	95	[243]	Fluorine	F	9	18.9984
Antimony	Sb	51	121.75	Francium	Fr	87	[223]
Argon	Ar	18	39.948	Gadolinium	Gd	64	157.25
Arsenic	As	33	74.9216	Gallium	Ga	31	69.72
Astatine	At	85	[210]	Germanium	Ge	32	72.59
Barium	Ba	56	137.34	Gold	Au	79	196.967
Berkelium	Bk	97	[247]	Hafnium	Hf	72	178.49
Beryllium	Be	4	9.0122	Helium	He	2	4.0026
Bismuth	Bi	83	208.980	Holmium	Ho	67	164.930
Boron	B	5	10.811	Hydrogen	H	1	1.00797
Bromine	Br	35	79.909	Indium	In	49	114.82
Cadmium	Cd	48	112.40	Iodine	I	53	126.9044
Calcium	Ca	20	40.08	Iridium	Ir	77	192.2
Californium	Cf	98	[251]	Iron	Fe	26	55.847
Carbon	C	6	12.01115	Krypton	Kr	36	83.80
Cerium	Ce	58	140.12	Lanthanum	La	57	138.91
Cesium	Cs	55	132.905	Lawrencium	Lw	103	[257]
Chlorine	Cl	17	35.453	Lead	Pb	82	207.19
Chromium	Cr	24	51.996	Lithium	Li	3	6.939
Cobalt	Co	27	58.9332	Lutetium	Lu	71	174.97
Copper	Cu	29	63.54	Magnesium	Mg	12	24.312
Curium	Cm	96	[247]	Manganese	Mn	25	54.9380
Dysprosium	Dy	66	162.50	Mendelevium	Md	101	[256]
Einsteinium	Es	99	[254]	Mercury	Hg	80	200.59
Erbium	Er	68	167.26	Molybdenum	Mo	42	95.94

* The atomic weights are based on the carbon-12 standard: $^{12}C = 12.00000$. Brackets denote a radioactive element; the number given is the mass number of the most abundant or most stable (longest half-life) known isotope of the element. (Elements 104 and 105 have not yet been officially named.)

Element	Symbol	Atomic Number	Atomic Weight	Element	Symbol	Atomic Number	Atomic Weight
Neodymium	Nd	60	144.24	Silicon	Si	14	28.086
Neon	Ne	10	20.183	Silver	Ag	47	107.870
Neptunium	Np	93	[237]	Sodium	Na	11	22.9898
Nickel	Ni	28	58.71	Strontium	Sr	38	87.62
Niobium	Nb	41	92.906	Sulfur	S	16	32.064
Nitrogen	N	7	14.0067	Tantalum	Ta	73	180.948
Nobelium	No	102	[254]	Technetium	Tc	43	[99]
Osmium	Os	76	190.2	Tellurium	Te	52	127.60
Oxygen	O	8	15.9994	Terbium	Tb	65	158.924
Palladium	Pd	46	106.4	Thallium	Tl	81	204.37
Phosphorus	P	15	30.9738	Thorium	Th	90	232.038
Platinum	Pt	78	195.09	Thulium	Tm	69	168.934
Plutonium	Pu	94	[244]	Tin	Sn	50	118.69
Polonium	Po	84	[210]	Titanium	Ti	22	47.90
Potassium	K	19	39.102	Tungsten	W	74	183.85
Praseodymium	Pr	59	140.907	Uranium	U	92	238.03
Promethium	Pm	61	[145]	Vanadium	V	23	50.942
Protactinium	Pa	91	[231]	Xenon	Xe	54	131.30
Radium	Ra	88	[226]	Ytterbium	Yb	70	173.04
Radon	Rn	86	[222]	Yttrium	Y	39	88.905
Rhenium	Re	75	186.2	Zinc	Zn	30	65.37
Rhodium	Rh	45	102.905	Zirconium	Zr	40	91.22
Rubidium	Rb	37	85.47	—	—	104	[260]
Ruthenium	Ru	44	101.07	—	—	105	[262]
Samarium	Sm	62	150.35				
Scandium	Sc	21	44.956				
Selenium	Se	34	78.96				

Table of Symbols

These symbols are used throughout the book. When both upper and lower case letters are shown for the same extensive property (one that depends upon system mass), upper case refers to a mole of material, lower case to a specific mass (e.g., gram or pound). When appropriate, dimensions are given in terms of m, l, t, and deg (for temperature).

a	constant (for a particular gas) in van der Waals equation of state reflecting attractive forces between molecules, l^5/t^2m; deceleration, l/t^2
b	covolume (volume occupied by molecular mass and thus inaccessible to other molecules) correction factor in equations of state, l^3/m
C	total heat capacity, ml^2/t^2deg
°C	Celsius temperature
C_p, c_p	heat capacity per mole or unit mass at constant pressure, l^2/t^2deg
C_V, c_v	heat capacity per mole or unit mass at constant volume, l^2/t^2deg
d	operator indicating exact or total differential change in what follows
E	total energy, ml^2/t^2
e	base for natural logarithms
F, f	force, ml/t^2
°F	Fahrenheit temperature
g	acceleration due to gravity, l/t^2
H, h	enthalpy per mole or unit mass, l^2/t^2
J	joule, ml^2/t^2
k	Boltzmann's constant, gas constant per molecule, l^2/t^2deg
k	kilo- (1000)
K	kelvin, unit of temperature on absolute scale
L	liter, l^3; sometimes length
l	length

m	mass
m	milli- (1/1000)
M	molecular weight
M	mega- (1,000,000)
n	number of moles; number of molecules/unit volume
N	Avogadro's number (number of molecules in a mole)
N	newton, ml/t^2
p	pressure, m/t^2l
P	probability
Q, q	heat interaction, ml^2/t^2
R	universal gas constant, $l^2/t^2\text{deg}$
R	Rankine, unit of temperature on absolute Fahrenheit scale; rarely, Réamur temperature
S, s	entropy per mole or unit mass, $l^2/t^2\text{deg}$
t	time; temperature on Celsius or Fahrenheit scale
U, u	internal energy per mole or unit mass, l^2/t^2; sometimes total internal energy, ml^2/t^2
v	velocity, l/t
V, v	volume per mole or unit mass, l^3/m; sometimes total volume, l^3
W, w	work, ml^2/t^2
x	arbitrary independent variable; sometimes distance, l
y	arbitrary dependent or independent variable
z	arbitrary dependent variable; sometimes height or altitude
γ	specific heat ratio, C_p/C_V or c_p/c_v
δ	operator indicating inexact or incomplete differential amount of what follows
∂	operator indicating a partial differential change in what follows, partial meaning that only one independent variable changes, any others remain constant.
Δ	operator indicating a finite change or difference in what follows
η	efficiency
θ	arbitrary empirical temperature, deg
π	ratio of the circumference of a circle to its diameter
ρ	mass density, m/l^3

Answers to Problems

1-3   3,300,000 ft lb

1-4   35,738 engines

2-2   4.22 K, $-452.1°F$, 7.61 R; 3683 K, 5678°F, 6138 R

2-3 -40

2-4   56, 10.7, 60.75, 68.5 (mm)

2-5   2.310; 631 K

2-6 $T_s = 85.83$; $T_{tp} = 62.83$; $t_{tp} = -62.83$

2-7   (a) 373.15; (b) 273.15; (c) 546.30; (d) $-99,999,354$

2-8   (a) $t = (\mathscr{E} - a)/(b + 100c)$;   (b) $\mathscr{E} = a + b(T - 273.15) + c(T - 273.15)^2$; $\mathscr{E} = a - 273.15b + (273.15)^2c$; (c) $\theta_{(t=-100)} = 200(1 - 100b/a + 10^4 c/a)$

2-9   (a) $T_s = 298.52$; $T_i = 218.52$; (b) -218.52; (c) 480.12

2-10 $T_i = 73.20$; $T_s = 100$; $T_{mp} = 160.83$

3-2   388 kg; 29.47 lbm mol

3-3   1.111 kg/m^3

3-4   A: 1.203 kg mol/m^3, 60.15 kg/m^3; B: 0.581 kg mol/m^3; 58.1 kg/m^3

3-5   rhodon > alphon > gammon

3-6   100.2

3-7   (a) 5.281×10^{18} kg; (b) 3 atm (std) or 2180 torr; (c) 413.6 N/m^2 or 3.103 torr

3-8   1937.2 kg

4-1   636.5 J

4-2 6.915×10^6 J; 4.988×10^6 J

4-3 3.099×10^6 J

4-4   (a) 6.235×10^6 J; (b) 8.644×10^6 J; (c) 150 K; 2.494×10^6 N/m^2

4-5 $(1.194 + b)$m^3

4-6 2.881×10^6 J; -1.858×10^6 J

4-7   2560 J

4-8   4304 balloons; 6.172×10^6 J

4-9   7890 ft lbf; 8291 ft lbf

4-10   5.44 watts

5-1 39.5°C

5-2 weak coffee at 35.2°C

5-3 0.146 cal/g °C

5-4 4.25×10^8 kg; 8.95°C

5-5 13.333 kg liquid and 1.666 kg vapor at 100°C

5-6 0.256 BTU/lbm °F; 0.256 cal/g K

5-7 18,872 kcal; some heat used to warm air that escaped

5-8 2235 K

5-9 $aT_1 - b \ln 2$

5-10 68.5°C

6-1 (b) 900 K; (c) 0.1 atm; (d) $2RT_1$ (1st step) vs. $6.9RT_1$ (2nd step)

6-2 (b): (i) $0.693RT_0$; (ii) 0; (iii) $-RT_0/4$; (iv) 0

6-3 900 K

6-4 twice as much

6-6 (b) $3T_1/2$; $3V_1/2$; (c) i and ii; (d) refrigerator because counterclockwise; also net work negative; (e) $-0.618RT_1$

6-7 678.6

7-1 16 ft/sec

7-2 43.2 ft/sec; less: 464 ft lbf vs. 640 ft lbf

7-3 1.63 times as fast or 118.8 ft/sec faster

7-4 5.87×10^4 g

7-5 0.221 gal

7-6 55 K

7-7 7.5 percent

7-8 4.376 J; 0.00104°C; 45 cm

8-1 $16,000R$; $20,000R$; 0 and $4000R$

8-2 (a) 1.559×10^7 J; (b) 3.118×10^6 J; (c) 3.602×10^5 J

8-3 (a) $2400R = 1.995 \times 10^4$ J; (b) 1500 K and 125 atm

8-4 No, because the claim is impossible—more power than HER allows.

8-5 (a) 1268 J; (b) 3.73

8-6 (a) 1.227×10^{12} J; (b) 2.250×10^9 g

8-7 (a) 8.48 watts; (b) 2¢ for power vs. $1.00 for ice

8-8 (a) 1034 m/sec; (b) 54,550 m

8-9 $C_V = BR$; $C_p = (B + 1)R$; $n = (B + 1)/B$

8-10 35 watts

8-11 (a) $p_b = p_0/2$, $p_c = p_0$; $V_b = V_c = 2V_0$; $T_b = T_0$; $T_c = 2T_0$ (c) $Q_{ab} = RT_0 \ln(1/2)$; $Q_{bc} = -2RT_0$; $Q_{ca} = 3RT_0$; $W_{ab} = RT_0 \ln(1/2)$; $W_{bc} = 0$; $W_{ca} = RT_0$; $W/Q_{abs} = 0.31$

8-12 36 L

10-1   26 percent

10-2 $-10.8°F$

10-3   17°C

10-4   13.3¢

10-5   35.3¢/gal

10-6   296.2 K

10-7   5696 cal/g mol; 6.145 atm

10-8   637 torr

10-9   21.15°C

11-1   13.5 mpg

11-2 $3075R$; 68 percent

11-3 $1939R$; 65 percent

11-4 $299R$; 7 percent (to 75 percent)

11-5 $184R$; 6 percent (to 71 percent)

11-6   (a) $4T_o$; (b) $2.81T_o$; (c) $-5.86T_o$; (d) $-0.953T_o$; efficiency is 77 percent

11-7   (b) $Q_{12} = RT_o \ln 4$; $Q_{23} = -3RT_o/2$; $Q_{34} = (-RT_o/2)\ln 4$; $Q_{41} = 3RT_o/2$ (c) efficiency is 24 percent

11-8   Carnot efficiency is 50 percent. Difference is due to irreversibility of heat transfers in constant volume steps. With recuperators, efficiency of Stirling cycle same as Carnot cycle.

12-2   (a) -1.287×10^{13} cal/K; (b) 3.2×10^{13} cal/K; (c) 1.912×10^{13} cal/K

12-3 2.27×10^7 kcal/K

12-4 6.666×10^8 J/K

12-5   1553 kg; $-19{,}200$ J/K, 1.91×10^6 J/K, 1.892×10^6 J/K

12-6   36.7 K; 4831 J/K, 5763 J/K, 4831 J/K (N.B. $\Delta S_{\text{reservoir}} = -5763$ J/K)

12-7 pV area is net work done; TS area is net heat converted to work.

12-9 $Q_{12} = \Delta S_{12} = 0$; $Q_{23} = 4800R$, $\Delta S_{23} = 3.90R$; $Q_{34} = \Delta S_{34} = 0$; $Q_{41} = 2400R$, $\Delta S_{41} = -3.90R$; $\Delta S_{\text{univ}} = 6.4R$

Index/Glossary

Boldface page numbers indicate main discussions or definitions.